INITIAL INTERVIEWING

What Students Want to Know

INITIAL INTERVIEWING
What Students Want to Know

FIRST EDITION

Tricia McClam
University of Tennessee, Knoxville

Marianne Woodside
University of Tennessee, Knoxville

The authors contributed equally to the writing of this book.

BROOKS/COLE
CENGAGE Learning™

Australia • Brazil • Japan • Korea • Mexico • Singapore • Spain • United Kingdom • United States

BROOKS/COLE
CENGAGE Learning™

Initial Interviewing:
**What Students Want to Know,
First Edition**

Tricia McClam and Marianne Woodside

Acquisitions Editor: Seth Dobrin

Assistant Editor: Allison Bowie

Editorial Assistant: Rachel McDonald

Technology Project Manager: Andrew Keay

Marketing Manager: Trent Whatcott

Marketing Assistant: Darlene Macanan

Marketing Communications Manager: Tami Strang

Content Project Manager: Rita Jaramillo

Creative Director: Rob Hugel

Art Director: Caryl Gorska

Print Buyer: Linda Hsu

Text Permissions Editor: Robert Kauser

Production Service: Seth Miller, Matrix Productions

Copy Editor: Ann Whetstone

Cover Designer: Dare Porter

Compositor: PrePress

For product information and technology assistance, contact us at
Cengage Learning Customer & Sales Support, 1-800-354-9706

For permission to use material from this text or product,
submit all requests online at **www.cengage.com/permissions.**
Further permissions questions can be emailed to
permissionrequest@cengage.com.

Library of Congress Control Number: 2008940672

ISBN-13: 978-0-495-50148-0

ISBN-10: 0-495-50148-4

Brooks/Cole
10 Davis Drive
Belmont, CA 94002-3098
USA

Cengage Learning is a leading provider of customized learning solutions with office locations around the globe, including Singapore, the United Kingdom, Australia, Mexico, Brazil, and Japan. Locate your local office at: **www.cengage.com/international.**

Cengage Learning products are represented in Canada by Nelson Education, Ltd.

To learn more about Brooks/Cole, visit **www.cengage.com/brookscole.** Purchase any of our products at your local college store or at our preferred online store **www.ichapters.com.**

Printed in the United States of America
1 2 3 4 5 6 7 12 11 10 09

CONTENTS

About the Authors

Tricia McClam

Tricia McClam is a professor in the Counselor Education program at the University of Tennessee, Knoxville. She teaches courses in the doctoral program, serves as Associate Head of the Department of Educational Psychology and Counseling, and coordinates the college's Grief Outreach Initiative. Her research broadly deals with international human services, service learning, and the helping process. She has published articles on these topics in various journals and has a record of national and international presentations at professional meetings. She is past editor of *International Education* and *Human Service Education*, both refereed journals. Included among her professional awards are the Helen B. Watson Outstanding Faculty Research Award in the College of Education, a Certificate of Appreciation from the Council for Standards in Human Service Education, and most recently, the Miriam Clubok Award from the National Organization for Human Service Education. Tricia McClam and Marianne Woodside have co-authored a number of books with Cengage Brooks/Cole.

Marianne R. Woodside

Dr. Woodside is affiliated with the Counselor Education program at the University of Tennessee, Knoxville. She currently coordinates the masters and doctoral programs in school, mental health, and counselor education. She teaches foundations of counselor education and supervises practicum and internships. Her primary research interests are international human services, counselor education and supervision, and the helping process. Dr. Woodside has served on the editorial boards of Human Service Education, Sage, and American

Association for Counseling and Development. She has published a wide range of articles in journals and has an extensive record of presentations at national and international meetings. Among her awards are the Distinguished Recognition Award from the Council for Standards in Human Service Education (2001) and the Professional Development Research Award from the National Organization for Human Services Education (1999). Marianne Woodside and Tricia McClam have co-authored a number of books with Cengage Brooks/Cole.

INTRODUCTION

Interviewing is an important tool in various settings—from the job interview to the first meeting with a new physician. In the helping professions, the interview, and interviewing, is vital. The initial or intake interview generally occurs at the beginning of service provision. The diagnostic interview identifies issues or possible disorders consistent with an established taxonomy like the *Diagnostic and Statistical Manual of Mental Disorders—TR Fourth Edition*. The mental status examination and the psychosocial interview determine the special needs of some clients. But regardless of the purpose of the interview, there is always a beginning encounter between the interviewer and the interviewee. We call this the *initial interview*.

The initial interview is the focus of this text. In six chapters we introduce the knowledge and skills needed for an effective first meeting between two people: one who is seeking help either voluntarily or involuntarily and one who will provide help. A successful initial interview is the foundation for a successful helping process, and the lessons taught in this text—the planning, relationship-building, skills, and strategies—can always be applied in the steps that follow.

The interviewee who comes for help may be any age, gender, religion, ethnicity, or race. At times, the interviewee may be more than one person, for example, a couple, a family, or a small group. Any helper may be an interviewer, including teachers; ministers, priests, or rabbis; social workers; counselors; and other human service professionals.

This text uses a question-and-answer format. Over a 10-year period, we collected questions from our students who were learning to be helping professionals. One of their first courses introduces interviewing and counseling techniques. At the beginning of each semester, students share their questions and concerns about interviewing. The result is this text. Because observing various professionals and interviewees will help you develop your own interviewing skills, a DVD accompanies the text to illustrate many of the interviewing concepts.

Each chapter introduces a critical component of interviewing. For example, Chapter 1 answers basic questions about interviewing—the nuts and bolts. The following five chapters then discuss in some detail the skills and strategies an effective interviewer needs.

Each chapter also includes three thought-provoking features: "Common but Difficult," "Cultural Context," and "Ethical Considerations." Any time you work with people, the potential exists for something to occur that is difficult to address. The purpose of "Common but Difficult" is to help you begin thinking about ways to respond to these situations. "Cultural Context" gives attention to diversity issues in interviewing. What can an interviewer do to maximize effectiveness with a diverse cliental? Expanding your awareness of your worldview and increasing your knowledge about a culture and its ramifications are our goals for this section. "Ethical Considerations" focuses on the ethical dimensions of interviewing. Some of the issues relate to the broader helping process as well, while others are unique to the interview setting. Awareness of ethical issues helps the interviewer develop sensitivity to professional issues and behavior that are so important to remember when working with clients.

Each chapter then concludes with opportunities for you to think about chapter content in terms of your own professional development. These "Personal Reflections" pose questions or situations related to the initial interview for your consideration.

As you read the brief responses to the questions generated by our students as they began their journeys as helping professionals, our hope is that you will delve further into the topics that intrigue you. The initial interview is a critical part of the helping process, the foundation for what follows. We hope these questions and answers are helpful to you as you begin your own journey.

Introduction to the Initial Interview

I remember my experience learning to conduct assessment interviews for individuals applying for Mobile Meal services. We were prepared to visit the homes of the applicants to determine if and why they needed this service and if the service would be short or long term, and to assess the financial need of the applicant. We needed to obtain personal information, emergency contact data, reasons for the request (current accessing of meals, accessing of meals on nondelivery days, day care information), criteria (e.g., living alone, unable to secure meals from other sources, inadequate ability to prepare meals, assistance from family or friends), details about medical conditions, level of mobility, and referrals to other agencies. We also provided the agency with our observations about the applicants and their living conditions. Before I conducted my first interview, I wondered how welcome I would be in applicants' homes. Would these individuals think I was invading their space and their privacy? How would I get the information I needed without being pushy or without appearing nosey? MW

Chapter 1 provides the first step on the path to becoming an effective interviewer. As Marianne shares in her brief description of assessment interview training, thinking about interviewing raises a number of questions and concerns. This is especially true if you are going into someone's home as she was. The type of information you are requesting can add to a beginning interviewer's discomfort by seeming like an invasion of privacy.

This chapter focuses on interviewing, especially the initial interview, as it relates to the process of helping and building the helping relationship. Questions like "What is an interview?" and "What is its purpose?" provide answers to basic questions about the first encounter between the two participants. This information then becomes the foundation for the knowledge and skills that are discussed in later chapters.

Why Do People Come for Help Today?

A variety of dilemmas and situations prompt individuals, families, or groups to seek help from a professional. Today's world is complicated, fast-paced, and often isolating. Many people live apart from their extended family and feel a lack of community. Support networks, if they exist, are very different today than they used to be. Television or the Internet may provide comfort, but most of the time they do not supply assistance for those who need it.

People experience difficulties in a number of areas. These include relationship problems at home, work, or leisure; problems created by environmental factors such as the economy, immigration, unemployment, or community situations; and personal issues such as illness, finances, and daily living. When we encounter a problem or difficulty we cannot solve ourselves, sometimes we feel more comfortable talking first with family, friends, or a trusted co-worker. At times, problems continue or they may represent complexities informal helpers do not know how to address. Then we may seek help from professionals such as a teacher; a social worker; a minister, priest, or rabbi; or a counselor. Let's look at several individuals who are experiencing difficulties.

> Tom has just lost his job. He is married and has three children. He owns a home and has credit card debt. His wife, Debby, is a waitperson at a local restaurant.
>
> Rohan is 8 years old and has been diagnosed with ADHD. He lives with his grandmother; his mother is incarcerated and no one knows where his father is.
>
> Pablo is a heroine addict; he has been clean for 16 days and just relapsed.

Tom, Rohan and his grandmother, and Pablo all represent individuals and families who need help. Their problems are complex and complicated, and may require support beyond that their friends or family can give—professional support.

How Do People Know Where to Go for Help?

There are several ways that people find professional help. Friends and co-workers may recommend an agency or a person that they themselves or a family member consulted in the past. The Yellow Pages of the telephone book offer a number of listings under headings such as counselors, psychologists, physicians, pastoral counselors, social workers, and social services. Many communities have published directories that target specific groups of people; for example, services for the elderly or for special education needs or listings of ongoing support groups. In today's competitive world, advertisements on the Internet, television, public buses, billboards, and flyers also alert people to available services. For those with immediate needs, hotlines and crisis centers are also resources. The World Wide Web (WWW) hosts websites of agencies, professionals, and organizations that provide help.

What Is the Difference Between a Helping Relationship and a Friendship?

At first glance, a friendship and a helping relationship may appear similar, perhaps because both encourage the sharing of problems and a strong sense of empathy. A helping relationship, however, is not a friendship. There are some significant differences. Helen Harris Perlman (1979), author of *Relationship: The Heart of Helping*, suggests some basic differences. First, a helping relationship is purposeful. It is established for an agreed upon purpose. The second is that it is time limited. Once the purpose or goal of helping is achieved, then the relationship is over. There is also a sense of authority on the part of the helper that comes from expertise in the knowledge and skills of helping. This is a bit different from our usual definition of authority. Friendships, on the other hand, do not have time limits, purpose, or authority. Friends are equals who have formed a relationship because of shared interests, concerns, or both. Often help is reciprocal in a friendship.

What Is the Helping Process?

Assistance to clients is provided by helping professionals as they engage in the helping process with clients. The helping process occurs in both formal and informal settings, and the interaction we talk about in this text occurs within formal settings. For example, many interviews take place in an office setting, but interviews may also occur where clients live or work, for example, in the home or on the streets. The effectiveness of the helping process depends greatly upon the skills of the helper, including the helper's ability to communicate an understanding of the client's feelings, clarify what the problem is, and provide appropriate assistance to resolve the problem.

The cornerstone of helping is the helping relationship. The helping process takes place within the context of a relationship that differs from others in that one person sets aside personal needs to focus on the needs of the other. The difference between a helper and a friend was described earlier. The helping process is grounded in an ethical orientation to helping others that includes attention to moral principles, ethical decision making, and attention and adherence to a professional code of ethics and legal standards.

The helping process includes preparation prior to meeting the client, exploring the problem, intervention strategies, and termination. Embedded within the helping process are two components: interviewing and change strategies (Nurius, Osborne, & Cormier, 2009). These two components of the helping process interface with each other throughout. Interviewing focuses more on relationship building, strength and problem assessment, and generating alternatives, and change focuses on implementation strategies. Although interviews are conducted with clients throughout the helping process, the focus of this text is on the initial interview that marks a time, in most instances, when the interviewee has come for help and/or advice (Stewart & Cash, 2002).

What Are Some of the Professional Organizations That Focus on Helping Others?

Most helping professionals are members of one or more professional organizations. Professional organizations promote standards of client care, detail professional and ethical behavior using a code of ethics, offer professional development, and expand the knowledge base through scholarship and research. Many professional organizations are listed below. If you would like more information, check the websites of these organizations.

American Association of Pastoral Counselors

American College Counseling Association

American Correctional Association

American Mental Health Counselors Association

American Rehabilitation Counseling Association

American School Counselors Association

Association for Adult Development and Aging

Association for Assessment in Counseling and Education

Association for Counselor Education and Supervision

Association for Gay Lesbian and Bisexual Issues in Counseling

Association for Multicultural Counseling and Development

Association for Specialists in Group Work

Association for Spiritual, Ethical, and Religious Values in Counseling

Council for Accreditation of Counseling and Related Educational Programs

Counselors for Social Justice

International Association of Addiction and Offender Counselors

International Association of Marriage and Family Counselor

National Association for Research and Therapy of Homosexuality

National Association of Social Workers

National Employment Counseling Association

National Organization for Human Services

National Board for Certified Counselors

How Does the Initial Interview Fit into the Helping Process?

The helping process usually begins with interviewing. It involves a minimum of two people who interact with each other. The interviewer brings knowledge, skills, values, and professional experience to the interaction. The interviewee also contributes knowledge, values, past experiences with helping, and strengths and problem(s) (see Table 1.1).

Generally, the interaction is face-to-face. Today, however, technology provides other vehicles for the interaction: cell phones, e-mail, and videos. These are

| TABLE 1.1 | WHAT INTERVIEWERS AND INTERVIEWEES BRING TO THE INTERVIEW PROCESS |

Interviewer	➡ Interview ⬅	Interviewee
Professional knowledge		Personal knowledge
Skills		Past experiences with helping
Value		Values
Experiences		Strengths and Problems

particularly helpful to those interviewees who live in remote or rural locations. In fact, in rural areas where there is no access to services, web communication, e-mail, and teleconferencing are used to communicate with interviewees during the initial interview.

Often the terms *interviewing* and *counseling* are used interchangeably. In fact, they are different. Interviewing is a basic foundational activity defined by its purpose of gathering information. Counseling may begin with an interview but it is a longer, more personal, and more intensive relationship between helper and helpee.

What Is an Initial Interview?

In more general terms, an interview is a formal meeting or a discussion that occurs around a particular goal or focus. As stated earlier, the interview is an integral part of the helping process. The helping interview is a special type of interaction between a helping professional and a person seeking help.

Within the context of helping, as discussed earlier, an initial interview is usually the first face-to-face contact between an interviewer and an interviewee. There may have been initial contacts by telephone or letter. If the interview occurs in response to a referral, the initial meeting should follow as soon as possible after the referral occurs. The interview is an opportunity for the interviewer and the interviewee to get to know one another; explore problems, situations, or both; and identify client resources and strengths.

Clarifying the reasons for the interview is often necessary. Is it to provide information, offer academic advising, explore career options, change behavior, cope with a diagnosis or disability, or acquire housing? Whatever its purpose, the interview is the beginning of the helping process and a time to build rapport, communicate understanding and acceptance, and talk about concerns and goals. As such, the information shared during the interview is important and requires an interviewer who is a skillful listener, interpreter, and questioner. These are skills that you will read about in later chapters.

EXERCISE 1.1: **Introduction to Interviewing**

Video Clip A introduces the DVD, its purpose, and the interviewers who will help you learn about the initial interview.

EXERCISE 1.2: **Some Thoughts on the Initial Interview: Video Clips B, C, and D**

On Video Clips B, C, and D, three professional helpers share some of their thoughts about an initial interview. They talk about their own experiences and how the context of their work determines the goals and focus of the initial interview.

Watch the video clips, and answer the following questions:

1. *What is the role of interviewing in their work?*
2. *How does each approach the task?*
3. *What questions do you have about their comments?*

How Long Is an Initial Interview?

Okun and Kantrowitz (2008) limit the use of the word *interview* to the first meeting, calling subsequent meetings *sessions*. In fact, the actual length of time of the initial meeting depends on a number of factors, including its purpose; the structure of the agency, organization, or school; the comprehensiveness of the services; the number of people involved (an individual, family, or small group); and the amount of necessary information.

Time varies for each interview and is often determined by its purpose, the caseload of the interviewer, the complexity of information needed, and the number of problems the interviewee brings. Many interviewers would like to have more time than they are allotted for the initial interviewer. A case manager from a community action agency recalled that she conducted intake interviews in 15 minutes. "When I was in school reading textbooks, I thought I would always have an hour and a half for every interview. And we would just ask questions to get all the information about this person. Not true!" (Janelle Stueck, Private Industry Council, personal communication).

Initial interviews vary. For instance, the initial interview with Rohan and his grandmother took place over two meetings. The helper first met with Rohan's grandmother to begin establishing a relationship with her and gathered social, medical, and psychological histories about Rohan. In the second initial interview, the helper met with Rohan during the school day. She conducted an informal session with him and used art therapy as a way to initiate a relationship with him and learn about how he perceived himself and his relationships at school and at home.

Where Does the Initial Interview Take Place?

Interviews generally take place in offices at agencies, schools, hospitals, and other institutional or agency settings. Sometimes, however, they are held in an interviewee's home. In such cases, the interviewer has the distinct advantage of observing the interviewee in the home, a situation that provides information that could not be gathered in an office setting. An informal location such as a park, a restaurant, or even the street can also serve as the scene of an interview. Whatever the setting, it is an important influence on the course of the interview. Note that in the case of Rohan and his grandmother, the initial interview involved two meetings, one at home and the other at school.

What Exactly Is the Structure of an Initial Interview?

Interviews often have three phases. The first phase can best be described as *contact and connection*. The interviewer and interviewee greet one another and begin to interact. Sharing thoughts, feelings, behaviors, events, and strengths comprise the middle of the interview. This is the *essence* of the interview. At the end, a summary provides a way to *close* by describing what has taken place during the interview and deciding what follows (Brill & Levine, 2002). A brief introduction to each phase follows. The skills relevant to each phase are described in the next chapters.

Contact and Connection

Several important activities occur at the beginning of the interview: greeting the interviewee, establishing the focus by discussing the purpose, clarifying roles, and exploring the problem that has precipitated the interview. The beginning is also an opportunity to respond to any questions that interviewees may have. They might want to know about your role, confidentiality, and the cost of the interview.

The Essence

During the middle phase of the interview, the interviewer focuses on developing his or her relationship with the interviewee. Any assessment, planning, and implementation also take place at this time. The type of assessment is defined by the specific purpose of the interview, often in accordance with the guidelines of an agency or organization. Sometimes assessment tends to be problem-focused, but a discussion that focuses only on the client's needs or weaknesses, or both, can be depressing and discouraging. Spending some of the interview identifying strengths can be energizing and result in a feeling of control. Assessment may also include consideration of the interviewee's eligibility for services in light of the information collected. All these activities lead to initial planning and implementation of subsequent steps, which may include additional data gathering or a follow-up appointment.

The Close

At the close, the interviewer and the interviewee have an opportunity to summarize what has occurred during the initial meeting. The summary of the interview brings this first contact to closure. Closure may take various forms, including the following scenarios.

- The interviewee asks for advice about what to do next.
- The interviewer provides direction for next steps.
- The interviewer and the interviewee believe that additional information from the interviewee and other sources may be needed before the interview can be completed.

What Types of Interviews Are There?

Because interviewing is used in multiple ways during the helping process, it is important to know how to conduct structured and unstructured interviews. The structured interview occurs when an agency or interviewer uses a standard set of questions for the interviewee. It may include demographic information such as name, age, address, family information, and other pertinent details about the interviewee. Information might include social history, brief medical history, financial information, and educational information. The information requested is very specific. For example, in gathering information about family history, specific questions might include "names of mother and father," "educational level of mother and father," "occupations of mother and father," "names and ages of siblings," "long-term relationship history" (relationship with long-term partner, number of years together), "educational level of long-term partner," and "occupational history of long-term partner." Usually this information is gathered early in the interview process and is considered intake information.

Table 1.2 shows an example of a structured initial interview. It illustrates the types of questions that a mental health specialist uses for each first interview with

---

TABLE 1.2 | STRUCTURED INTERVIEW

Date of Initial Interview _____
Name of Interviewer _____
Name of Client _____
Age _____   Gender _____   Ethnicity _____   Primary language _____

1. The identified problem as described by the interviewee
2. Prior counseling and focus
3. Explain the issues as you see them.
4. What do you think that I should know about these issues?
5. List all of the problems you are experiencing?
6. Which of these issues are most important for you to solve? On a scale of 1 – 4 list the severity of these problems. (1 not serious; 2 not very serious; 3 serious; 4 very serious)
7. How long have you been having these issues?
8. Why do you think that these issues exist?
9. Who else knows about these issues? What do these individuals say about them?
10. Why it is difficult to deal with the issues?
11. How have you tried to solve the issues?
12. What problems have you been successful solving?
13. What did you do to solve them successfully?
14. What do you think your skills are?
15. What type of support do you have?
16. What kind of assistance would you like?
17. What type of changes would you like to see?

a client. In this interview format, the helper structures the conversation to address both issues and interviewee strengths.

An unstructured interview typically includes a number of areas of focus for the interview, such as social history, educational information, and family history, but the questions are more open-ended in nature. For example, an interviewer might ask the interviewee, "Tell me about your family." The interviewee would determine what he would talk about; the interviewer would encourage him to talk in depth by asking, "Can you tell me more about that?" or "Do you have any specific examples you could share?"

In an interview with even less structure, the interviewee determines the areas of self-disclosure. The interviewer begins with a general inquiry: "Tell me about you." Or "Let's talk about why you're here." In this type of interview, it is important for the interviewee to focus the discussion. The interviewer's role is to encourage the discussion of issues as fully as possible: "Tell me more about that," "Let's talk about this," or "Give me an example."

The motivational interview (Miller & Rollnick, 2002) focuses on the readiness of the interviewee to change. In this type of interview, the interviewer's goal is to help the interviewee see issues and problems and their consequences more clearly and then help him see a better life once the problems and issues are addressed. The motivational interview uses empathy to establish relationships and encourages self-efficacy. Even though the work is called *interviewing*, the interviewing or beginning of this work is part of a broader therapeutic technique.

 EXERCISE 1.3: **Types of Interviews**

In *Video Clips E and F a professional helper conducts a structured and unstructured interview with the same interviewee Watch Videos E and F, and answer the following questions*:

1. *What are the major differences that you note between structured and unstructured interviews?*
2. *Describe the strengths and limitations of each type of interview.*

What Do All Helping Interviews Have in Common?

There are commonalities in all interviews in the helping professions that should occur in any initial interview. First, there must be shared or mutual interaction: Communication between the two participants is established, and both share information. The interviewer may be sharing information about the agency and its services, while the interviewee may be describing the problem. No matter what the subject of their conversation is, the two participants are clearly engaged as they develop a relationship.

Second, the participants in the interview are interdependent and influence each other. Each comes to the interaction with attitudes, values, beliefs, and experiences. As we stated earlier in the chapter, the interviewer also brings the knowledge and skills of helping, while the person seeking help brings personal strengths and the problem(s) that is causing distress. As the relationship develops, whatever one participant says or feels triggers a response in the other participant, who then shares that response. This type of exchange builds the relationship through the sharing of information, feelings, and reactions.

The third commonality is the interviewing skill of the interviewer. She remains in control of the interaction and clearly sets the tone for what is taking place. Knowledge and expertise of interviewing process distinguishes the interviewer from the interviewee and from any informal helpers who have previously been consulted. Because the helping relationship develops for a specific purpose and often has time constraints, it is important for the interviewer to bring these characteristics to the interaction, in addition to providing information about the agency, its services, the eligibility criteria, community resources, and so forth.

Personal Reflections

1. *As indicated earlier, a helping relationship with a professional and a relationship with a friend are very different. Describe a relationship you have had with each, and note the differences.*
2. *You have read about the different types of initial interviews used by helping professionals. If you were the interviewer, which type of interview would you prefer to use in an initial interview? Why? If you were an interviewee, which type of interview would you prefer to engage in? Why?*

REFERENCES

Brill, N. I., & Levine, J. (2002). *Working with people: The helping process* (7th ed.). Boston: Allyn & Bacon.

Ivey, A. E., & Ivey, M. B. (2008). *Essentials of intentional interviewing: Counseling in a multicultural world.* Pacific Grove, CA: Brooks/Cole.

Miller, W. R., & Rollnick, S. (2002). *Motivational interviewing: Preparing people to change.* New York: Guilford Press.

Nurius, P. S., Osborne, C. J., & Cormier, L. S. (2009). *Interviewing and change strategies for helpers: Fundamental skills and cognitive behavioral interventions* (6th ed.). Pacific Grove, CA: Brooks/Cole.

Okun, B. F., & Kantrowitz, R. E. (2008). *Effective helping: Interviewing and counseling techniques* (7th ed.). Pacific Grove, CA: Brooks/Cole.

Perlman, H. H. (1979). *Relationship: The heart of helping.* Chicago: University of Chicago Press.

Stewart, C. J., & Cash, W. B. (2002). *Interviewing: Principles and practices.* Boston: McGraw-Hill.

CONTACT AND CONNECTION | CHAPTER 2

I can still remember my first office. It was one of four identical offices all in a row. Each office was only large enough for a desk and two chairs. All offices had one solid wall that was behind the employee. Everything was institutional green. The other three walls were about two-thirds glass so there was no confidentiality, particularly when we began working with clients who were deaf and used American Sign Language to communicate. On the other hand, I was really glad that the glass walls were there when I worked with clients who were unpredictable, had a history of violence, or had a medical condition that might include seizures. TM

As Tricia describes her first office and its positive and negative features, it is clear the 8-foot by 8-foot space offered little flexibility. But the arrangement of an office is only one of the ways we make contact and connect with interviewees. This and other strategies to maximize the results of the initial meeting between interviewer and interviewee are the focus of Chapter 2. Regardless of setting, helping begins with the first contact between the person seeking assistance and a service provider that results in a *connection* or relationship. This contact begins within the context of the initial interview. The other strategies described in this chapter help you with tasks that begin the first phase of an interview. They lay the foundation for establishing a relationship. For example, how you arrange your office, greet the individual, ask for information, and listen all play a part in *connecting*.

Why Is the First Contact with the Interviewee so Important?

Both participants in an interview have expectations for the first meeting. Interviewers may or may not know why a parent, a senior, an adolescent, a family, or a small group is coming. Although we might like to assume people are coming voluntarily and will be cooperative and motivated, this is not always the case. So it is incumbent on you as the interviewer to approach the initial meeting with an open mind. This means reviewing any available information, preparing for the meeting by making sure disruptions are minimal, and taking steps to insure privacy.

One caution: It is important for the interviewer to gather initial impressions about individuals without making snap judgments (Todonov, Baron, & Oosterhof, 2008). Each day, we have more information coming in than we can process. In order to handle the influx, we take short cuts in our information processing and make judgments automatically (Weiten, Lloyd, Dunn, & Hammer, 2008). For example, if an individual comes in dirty and disheveled, the interviewer might make one of the following assumptions: "This person does not care how he or she looks," "This person has a job performing manual labor," or "This interviewee does not believe that this meeting is important." Any of these assumptions may be true; it also may be that the interviewee stopped on the way to the appointment to help someone whose car had broken down! Interviewers are going to have first impressions. It is important to record them, be aware that they are limited, and move beyond them to understand the interviewee in more depth.

The interviewee may have different expectations for the initial contact. This person may expect an interviewer to be prompt, competent, accessible, and non-judgmental, and may or may not be fearful or suspicious of the interviewer. This initial interaction is critical if the interviewee is to leave believing that the interviewer understood the problem and can help to resolve it. Factors that may contribute to a positive impression include the interview setting, the interviewer's appearance, and the manner in which the interview is conducted. We will provide more on these points later.

How Does the Interviewee View the First Meeting?

We believe that the interviewee views the first contact of the interview differently from the interviewer. For most interviews (except, for example, those that occur in the interviewee's home), interviewees approach the event as a stranger. They are entering a world about which they may know little. Usually interviewees make an appointment because they know they need help or someone has talked them into going for help. Many times even thinking about talking about problems and feelings to a stranger creates anxiety. During the initial interview, the interviewee may feel stressed, reluctant, or both.

Arranging the initial interview might also be difficult for the interviewee. He or she might have to arrange time from work, locate child care, or even bring a child, parent, or friend to the interview. Arranging transportation, finding the interview location, and walking into a building with little knowledge of what to expect can be difficult and intimidating.

Interviewing children has its own complexities. From the child's perspective, going to see the school counselor is out of the ordinary school day experience. Although the counselor is welcoming, the child is unsure of the interview situation.

How Important Is the First Impression?

First impressions occur in the first three seconds as interviewees assess your appearance and behavior (Gladwell, 2005). In this short time frame, your demeanor, dress, mannerisms, and body language are scrutinized. Grooming, accessories, handbag or briefcase, and jewelry may "speak" for you. Perceptions of the interviewer are influenced by these factors. This first impression remains with the interviewee unless she or he makes an effort to change it.

Sometimes you may intend to convey one message, but the interviewee may receive a different message. For example, you decide not to take notes during the interview so you can focus on the interviewee. The interviewee may respond positively and appreciate your undivided attention; or the interviewee may see your lack of note taking as inattention to detail and may wonder how you could possibility remember all that was said during the interview. During an interview, an interviewer wants to appear thoughtful so he/she doesn't smile much during the interview; the interviewee believes that you are distant and uninterested. From these examples, it is clear that your first impressions of appearance and behavior can affect the relationship.

Making a positive first impression works for the interviewer. These first few seconds start the interaction in a positive direction; otherwise, establishing a connection may become difficult. Communicating your accessibility, interest, positive orientation, and enthusiasm about the meeting and reinforcing the interviewee's presence are ways to make the first impression a positive one.

How Can I Make a Good Initial First Impression?

There are several ways that you can make a positive connection in the first few minutes of the interview. Signal to the interviewee that he or she is important by being punctual. All of us have been on time for appointments only to have to wait to see teachers, physicians, and other professionals. Beginning on time expresses to the interviewee that "you are important to me, and I respect your time and your effort to be here promptly" (Mullins, 1995).

Did you know that approximately 18 different smiles have been identified (Ekman, 1985; Ekman & O'Sullivan, 1991)? For example, a smile of enjoyment involves both the lips and the muscles around the eyes. The polite, social smile involves only the lips and usually only on one side, while the embarrassed smile causes the round part of the chin to move up and the face turns away briefly. Be sure your smile communicates a warm, open, genuine greeting. The interviewee will know if it doesn't.

The interviewee will also know if you are at ease and comfortable in the interview environment (Bedi, 2006). Even though you are just learning how to interview, if you can relax and be "in the moment," the interviewee will take a cue from you and relax, too. Prior to greeting the interviewee, you may use breathing techniques or stretches, or review the files (Weiten et al., 2008). It is important that you find a way to convey your ease with the interview process.

Your dress depends on the culture of your setting as well as the tone you wish to establish for the interview. For example, one agency we know is very specific about the dress code, listing as unacceptable such things as flip flops, visible body piercings and tattoos, halter tops, and plunging necklines. This particular agency does allow head coverings required for religious purposes. Another organization that serves children ages 6–10 suggests that professionals wear casual clothes, including jeans and t-shirts tucked in. So dress codes will vary. Whatever the code, neatness and cleanliness communicate respect for one's self and for others.

In conclusion, your initial presentation is important, so you need to look professional. What does that mean? First, it is different for each office. Look around and see what your colleagues are wearing. Take cues from them and dress appropriately. Ask your supervisor if you are unsure about dress. If you are interviewing children, you may need to wear clothing that allows you to talk and play with them; you may be kneeling or sitting on the floor. On the other hand, you might conduct interviews at court and need to dress up for court appearances. Neat and clean is always appropriate!

Does the Interview Begin When the Client Walks in the Door?

Many people think this means when the interviewee appears at the door, but it actually begins before then. The first contact may be making an appointment by phone or sending a letter and information about service delivery. Another way the interview begins is an initial phone conversation with a colleague who is making a referral. It is important for you to remember that for the interviewer, the interview begins before the interviewee arrives.

In preparing for the interview, most interviewers think about the goals for the first meeting, read over any information they may have about the individual, think about how to arrange the setting to facilitate communication and establish rapport, and determine how to plan for meeting the identified needs. The interviewer is also aware of any information needed before the interview or necessary forms that need to be completed during the interview. Searching for the form is distracting for both the interviewer and the interviewee.

How Do I Arrange the Interview Setting to Facilitate Relationship Building?

Interviewing can occur anywhere: an office, a bedside, a school, a local fast food restaurant, a home, or even the sidewalk. Today, some corporations even conduct interviews in airports to reduce the cost of bringing a number of applicants to a distant location. In international education and business, interviews may occur via conference calls, video conferencing, or the Internet.

Environment is important, however, and may actually have healing effects, so it is worthy of our consideration as we think about interviewing. In many cases there may be aspects of the environment that the interviewer can control and actually use to enhance the interview and the building of a helping relationship. For example, soft lighting or natural light may facilitate self-disclosure. Color is one of the most powerful aspects of the environment. Blue is a favorite color of both sexes and all ages. Darker colors make larger offices more intimate while white and very light colors make a small room larger. Light colors also are associated with positive emotions by children and young adults (Pressly & Heesacker, 2001). Considering the age and gender of one's clientele when choosing colors is always a good idea.

Interviewees receive messages from the environment, and accessories provide some powerful messages because they symbolize ownership of a space (Bedi, 2006). They also can make the helping environment more appealing. Plants, for example, are beneficial, particularly to aging individuals or those who are terminally ill because they represent life and growth. Another example is artwork; most appealing are pictures of animals, valleys, mountains, or farmland. On the other hand, most people don't care to look at artwork they don't understand. Personal items, for example, pictures of children or family, are also helpful in making the environment attractive and welcoming.

It is also the case, however, that an interviewer may not have control over the environment. Furniture may be modular or the walls painted institutional green, as Tricia experienced at her first job. In these situations, the most the interviewer can do is make sure the interviewee is seated in such a way that distractions are minimal. For example, facing a window or door may focus attention away from the interview.

We believe the most productive setting is one in which the interviewer and the interviewee are seated in such a way that barriers are minimal, including crossed arms or legs, and eye contact is maintained, if appropriate. It is also important that the location be accessible to everyone, including those with disabilities, and soundproof to insure confidentiality. Where environments can be controlled, the interviewer should pay attention to factors such as lighting and accessories.

What Plans Should I Make for the Interview?

It is helpful to think about an outline of the *ideal* initial interview. Planning for this interview might include the following: (1) gathering agency materials, (2) arranging the room, (3) outlining the session, and (4) planning the contents of a summary of a session and next steps. An outline of a session might include the greeting, information about the interviewer, discussion of confidentiality and its limitations, and the purpose of the meeting. The interview then moves to clarifying expectations, gathering information, making an initial assessment, and determining what follows.

For a beginning interviewer, an outline of the initial interview provides a structure for the session that guides the interviewer. As the interviewer gains experience, adjustments to the structure occur based on a reduction of initial anxiety and a familiarity with interviewing.

EXERCISE 2.1: Beginning the Interview

In Video Clip A you will see a licensed mental health counselor preparing to interview a couple. Listen as the counselor talks about the office and prepares for the interview. Two questions guide your observations:

1. *How does the counselor talk about arranging the office to facilitate relationship building?*
2. *What does counselor do to get ready for the interview?*

Should I Establish Goals for the Interview?

The initial interview is a purposeful activity. As such, interview goals vary depending on its purpose. For example, if you are interviewing for an agency or organization that prescribes a number of questions during the first interview, then the mission and goals may determine the data you need to gather. If the setting allows a more open-ended first interview, your primary goal may be to determine the purpose of the meeting and establish rapport with the interviewee. Generally, initial interviews have three primary goals: (1) establishing rapport; (2) gathering data, and (3) making an initial assessment. We believe that it is helpful to have at least a general outline or goal in mind for the interview; we also think that interviewers need to provide some time in the interview for the interviewee to ask questions and voice any needs or concerns.

How Do I Greet the Interviewee?

The tone of the interview is set at the first contact. We have found that standing, smiling, and a handshake establish a positive tone right off the bat. These behaviors are important because it is estimated that between 65 and 93 percent of the social meaning in a normal conversation is nonverbal (Ambady & Rosenthal, 1992; Ketrow, 1999). So these gestures are important communications that begin the establishment of a positive relationship. Let's examine each of these behaviors.

Standing when the interviewee enters communicates respect and equality. Moving from behind a desk also reduces barriers, further emphasizing equality and respect. Standing is appropriate regardless of race, ethnicity, or gender.

The handshake is a bit more complicated because it may be uncomfortable for some, not acceptable culturally, or both. Touch has a powerful impact, often more powerful than words. Its meaning will be determined by what part of the body is touched, how long the touch lasts, the strength of the touch, and the method of touching. We evaluate touch in relation to early experiences with it. In some cultures, a bow is more acceptable than a handshake while in others, touch between men and women is unacceptable. During a recent visit to the Middle East, Tricia, in her friendly American way, extended her hand to an Islamic priest who put his hand behind his back and shook his head "no." There will be some unknown in the greeting as to whether or not to shake hands. Our best advice here is that if you think a handshake is questionable, take your cue from the interviewee.

What About Introductions?

It's important to remember that you are in charge. Usually this initial meeting occurs on your turf, whether in a school, office, hospital, or some other setting. It's up to you to take the lead. Introduce yourself, and let the client know what to call you: "I'm Joe Blow. Please call me Joe." or "I'm Dr. Blow." And if the interviewee doesn't volunteer a name or express a preference for what he or she would like to be called, then ask: "And you are?" or "What would you like me to call you?'

Pay close attention as the interviewee says his or her name. It's important that you not only remember the name but also pronounce it correctly. This is one way that we communicate that the interviewee is important and has our complete attention.

"I'm Jane Smith. Please call me Jane."

"I'm Keanu Reeves."

"I'm glad to see you. Please sit, Keanu (gestures to a chair). Did I pronounce your name correctly?"

"Yes, you did."

"I don't believe I've heard that name before. Is it a family name?"

In this example, Jane is using an inquiry about the first name for several reasons. This is one way to reinforce the correct pronunciation, break the ice, and most importantly, express genuine interest in this individual.

Another thing that happens in this example is Jane indicates where she would like Keanu to sit. Because her office is small, there is only one chair, and she gestures toward it. Larger offices may present more seating choices. Some interviewers believe it is significant where couples or family members sit. Who shares the sofa or who sits beside whom? In any case you'll want to be aware of where interviewees sit when there is a choice.

What Should I Share About Myself?

Part of the introduction is explaining who you are and the purpose of the interview. What are your credentials? Is there a plan? How much time will you spend together? What will we do? One source that informs your introduction is your profession's code of ethics. You will want to be familiar with the guidance it provides.

Basically, professional disclosure statements introduce you and may include basic information about you and your role. For example, your job title, any areas of expertise, licensure or certification, education and training, and length of time at the agency or office are often part of this introduction and shared with the interviewee.

Let me tell you a little about myself. I am a licensed professional counselor in this state. This means I have a master's degree in counseling and supervised counseling experience, and I have successfully passed the state test. I have been at this agency for four years. I know there is information you would like about our services like hours and fees. This brochure provides answers to our most frequently asked questions. If you have other questions or if I can explain these further, please let me know.

Sometimes this information about you is written in a brochure or other agency literature. In fact, the American Counseling Association Code of Ethics and Standards of Practice states that "counselors must adequately inform clients, preferably in writing, regarding the counseling process and counseling relationship at or before the time it begins and throughout the relationship." Specifically, A.3.a lists "purposes, goals, techniques, procedures, limitations, potential risks, and benefits of services to be performed." It is shared verbally or written in a brochure or other agency literature.

Linda, one counselor we know, says this about her practice's website:

> We put a lot of thought into how we described ourselves and our services on our website and in our agency's brochure. We debated a long time about whether to include "Christian" in our name, given the fact that it means so many different things to people, some of them not so good. We thought it was important, however, to include it as part of our effort to abide by the principle of full disclosure.

When Linda meets clients for the first time, she makes sure they know that she is a counselor whose worldview is informed by her Christian faith, and that she believes sound psychological principles and Biblical principles are in agreement. It's important to her that prospective clients know her worldview because the relationship is an important part of the healing process. She also believes that Christian counselors should be trained and licensed in accordance with state laws and regulations. She points out that none of us would go to a Christian physician who hadn't been to medical school or to a Christian lawyer who hadn't passed the bar exam!

It is up to you to determine how much information you share with the interviewee. Certainly you don't want to bore the interviewee or provide more information than can be processed. And you want to be guided by your profession's code of ethics. Remember the focus of the meeting is the interviewee, not you. As such, encourage the interviewee to share any questions or concerns about you, the helping process, or both.

How Do I Break the Ice?

Comments or questions at the beginning of the interview that are unrelated to the purpose of the interview are often called *icebreakers* or *door openers*. These are nonthreatening, casual comments that give the interviewee an opportunity to feel more comfortable by looking around the room, getting a good look at you, and becoming comfortable in the seat. The weather is usually a good topic, particularly if it's significant in some way—good, bad, hot, or rainy. Or you might ask if the interviewee had any trouble parking or finding the office. One social worker we know commented on an adolescent's t-shirt that depicted a rock group that had performed the night before; he reported that it was instant rapport!

EXERCISE 2.2: **Greeting the Client**

In Video Clip B, the counselor demonstrates how to begin an initial interview with a couple. Watch Video B, and answer these questions:

1. *Describe what happens as the couple enters.*
2. *What did you learn about the counselor and the agency?*
3. *How did the counselor put the couple at ease?*
4. *What do you think was most effective in this video clip?*

COMMON BUT DIFFICULT: **Recognizing Language Barriers**

SCENARIO A: Your next interview is with a woman who is deaf. You do not know American Sign Language (ASL), and your agency has no interpreters. You begin to imagine the difficulties of interviewing by writing notes to each other. But wait! The interviewee's 11-year-old daughter plans to accompany her mother and has interpreted for her before.

SCENARIO B: Your first interview after lunch is with a 30-year-old male from a village in central Mexico. He speaks little English and your high school Spanish is all but forgotten. In addition, you are puzzled by the Spanish he speaks. It doesn't sound like your high school Spanish.

In both of these scenarios, communication is a problem. No common language exists between the interviewer and the interviewee. This is a major barrier. In Scenario A, the daughter can help, but this raises serious questions about confidentiality. In Scenario B, the potential for misunderstanding exists since neither speaks the same dialect.

These two scenarios illustrate the need for an interpreter. Rather than risk a frustrating and confusing encounter for both participants that will accomplish little, one strategy is to reschedule the meeting when an interpreter can be present.

Have you ever worked with an interpreter? If not, what do you need to know? First, interpreters are highly trained professionals who are trustworthy, maintain confidentiality, and facilitate communication between or among the primary participants in an interview. To learn more about interpreters, their professional conduct, and their use, read "Working with Sign Language Interpreters in Human Service Settings," by Dr. Jeffrey Davis, who is an interpreter himself. You will find this article on your DVD. It explains how to work with an interpreter and what you might expect from one. Many of the principles you will read about are also true for interpreters of foreign languages such as Spanish or Swahili.

CULTURAL CONTEXT: **The Initial Greeting**

As indicated in this chapter, the way in which the interviewer greets the interviewee is important and contributes to the tone or tenor of the interview. The greeting is even more important if the interviewer and the interviewee come from different cultures (Sue & Sue, 2007). The interviewer contributes to the interview in two positive ways when greeting the interviewee in terms of his or her own customs. First, there are many greetings that the interviewer may use in his or her own culture that violate the cultural norms of the interviewee. Second, getting the greeting "right" sends a message to the interviewee, "I care about you and your culture and respect its ways." For example, for many Muslim women, touching individuals of the opposite sex violates their religious and cultural values. In fact, they are forbidden to do so. So what would an interviewer do, in this situation, to greet the client? An interviewer greeting a Muslim woman might open the door, nod the head or bow slightly, to acknowledge the presence of the interviewee. Then the interviewer could give a hand signal that beckons the interviewee in to the office and another hand signal that shows the interviewee a place to sit, with accompanying words such as, "I am glad that you came today. Won't you please sit down?"

ETHICAL CONSIDERATIONS: **Confidentiality**

Confidentiality is one of the most important topics you will discuss with the interviewee. The practice of confidentiality represents the integration of professional ethics and legal mandates and centers around the right to privacy. In the interview, trust and confidence provide a foundation for the work that occurs. Many interviewees will want to distinguish between the information that they give you and the information you are required to enter into the official record.

One of the first responsibilities of the interviewer is to inform the interviewee of the limitations of confidentiality (Mitchell, Disque, & Robertson, 2002). The Code of Ethical Principles for Marriage and Family Therapists specifically identifies the conditions under which they will disclose client confidences: "(1) mandated by law; (2) to prevent a clear and immediate danger to a person or persons; (3) where the marriage and family therapist is a defendant in a civil, criminal or disciplinary action arising from the therapy; or if there is a waiver previously obtained in writing…" (2.1). You should be familiar with the sections of your profession's ethical code that pertain to clients' rights and confidentiality.

As a way of respecting the confidentiality of the interviewee, focus on the client's written record or file. The interviewer can continually clarify what information will be part of the permanent file. The following are other confidentiality matters the interviewer can share with the interviewee:

- Explain what information is typically shared in interoffice staffings.
- Provide information about the Health Insurance Portability and Accountability Act (HIPAA) regulations the office or agency follows.
- Assure the interviewee that if information is to be shared with others outside the agency or office, permission from her will be sought. Rather than share an entire case file, the interviewer will share only the information that is needed.
- Assure the interviewee if information is needed from another agency, permission from the interviewee will be sought.

Finally, as mentioned earlier in this chapter, it is important for the interview to occur in a confidential location. Attending to this will help the interviewee feel more confident about sharing personal information, and the interviewer will be fulfilling ethical obligations.

Personal Reflections

1. *Think about a time you were interviewed. Describe the experience. Was it positive? Negative? Difficult? Enjoyable? Explain.*

2. *Think about a time when you wanted to make a good impression. How did you prepare for the encounter?*

3. *Have you ever had the opportunity to interact with someone from another culture? Briefly describe the experience. What was it like for you?*

REFERENCES

Ambady, N., & Rosenthal, R. (1992). The slices of expressive behavior as predictors of interpersonal consequences; A meta-analysis. *Psychological Bulletin, 111*, 256–274.

Bedi, R. P. (2006). Concept mapping the client's perspective on counseling alliance formation. *Journal of Counseling Psychology, 53*(1), 26–35.

Dayan, Y., (2008). Interviewing young children. *Exchange, 19*, 54–57.

Ekman, P. (1985). *Telling lies*. New York: Norton.

Ekman, P., & O'Sullivan, M. (1991). Facial expression: Methods, means, and moves. In Robert S. Feldman and Bernard Rime (Eds.), *Fundamentals of nonverbal behavior*, pp. 163–199. Cambridge: Cambridge University Press.

Gladwell, M. (2005). *Blink: The power of thinking without thinking*. New York: Little Brown, & Company.

Ketrow, S. (1999). Nonverbal aspects of group communication. In L. Frey (Ed.), *Handbook of group communication theory and research*, pp. 251–287. Thousand Oaks, CA: Sage.

Mitchell, C. W., Disque, J. G., & Robertson, P. (2002). What parents want to know. *Professional School Counseling, 6*, 156–161.

Mullins, J. J. Conveying messages that inspire action. *The Instrumentalist, 49*, 25–28.

Pressly, P. K., & Heesacker, M. (2001). The physical environment and counseling: A review of theory and research. *Journal of Counseling & Development, 79*, 148–160.

Sue, D. W., & Sue, D. (2007). *Counseling the culturally diverse: Theory and practice* (5th ed.). New York: John Wiley.

Terry, M. (2003). Best practices in interviewing. *Adult Learning, 14*(4), 26–27.

Todonov, A., Baron, S. G., & Oosterhof, N. N. (2008). Evaluating face trustworthiness: A model based approach. *Social Cognitive and Affective Neuroscience, 3*(2), 119–127.

Weiten, W., Lloyd, M. A., Dunn, D. S., & Hammer, Y. (2008). *Psychology applied to modern life: Adjustment in the 21st century* (9th ed.). Pacific Grove, CA: Wadsworth.

3 | # RAPPORT AND RELATIONSHIP

Early in my career, I was an elementary school counselor in Appalachia. One of my primary responsibilities was to visit the homes of students with whom I was working. One first visit stands out vividly. I drove from a paved road to a gravel road, and then to a dirt road. Parking my car at the end of the lane, I walked about a half mile to my student's home. I had prepared myself for this initial interview by considering my dress (not too dressy, but professional, sensible walking shoes, no purse), my professional papers (a small notebook and a pen with my notes on the purpose of the visit), and my introduction (friendly, thanking them for letting me come, and a short handshake and introduction). Nothing prepared me for the 300-square-foot home that I entered. It had no screens, a dirt floor, and no telephone. The visit went fairly well since I was delivering good news about their son. I left their home and walked back to my car only to discover it would not start. The relationship was strengthened when I returned to their home to ask for help. We bonded when I needed their help, and the relationship became more reciprocal in nature. MW

The relationship between the interviewer and the interviewee is the foundation of the helping relationship. Marianne was aware of this as she prepared for this home visit by thinking about the things *she* could do to contribute to a positive relationship with this family. When the interviewer establishes rapport with the interviewee, there is an increased likelihood that the goals of the initial interview can be achieved. This chapter explores the nature of rapport and answers questions that relate to establishing a relationship during the interview.

Why Build a Relationship?

The helping relationship is the vehicle through which help is offered. The relationship is actually more important than the helper's mastery of counseling theories (Brill & Levine, 2004; Cochran & Cochran, 2006; Okun & Kantrowitz, 2008). The thinking is that the relationship provides the foundation for what follows. Without a helping relationship built on respect, trust, and acceptance, helping strategies are much less effective. Rapport based on these qualities promotes the sharing of personal and often painful experiences.

What Is Rapport?

The origin of the word *rapport* comes from the Old French word *rapoter*, meaning to bring back. The Latin roots are *re* meaning back or again and *aporter*, meaning *to bring* (*American Heritage Dictionary, 2000*). What does it mean to establish rapport with someone? Interviewers who build rapport indicate that they understand interviewees by entering their world. Interviewees believe you are then "in sync," "on the same page," or "whistling the same tune." This means that we are accepting our interviewees for who they are: this includes their feelings, beliefs, and attitudes. As we convey to them, "I am with you in the moment," they may feel safe and be willing to explore the difficulty they are facing.

How Does an Interviewer Achieve Rapport with Another Person?

It is not enough to say that you understand the interviewee's situation. You have to demonstrate that you grasp his perspective or worldview. One way to accomplish this is to put aside your own thoughts in order to truly listen to the interviewee. As you listen, you try to suspend any judgments or thoughts you have about what the interviewee is saying so that you can fully attend to the interaction without thinking about what you want to say next. In this way you can enter into an interviewee's world.

Responses to the client range from the simple to the sophisticated. One way to begin is with nonverbal communication such as maintaining eye contact. Some interviewers try to mirror the interviewees' posture. Examples include sitting up straight or leaning forward, folding arms, tilting the head, all in alignment with the interviewees' body position. It is helpful to use verbal responses that indicate you are trying to understand the interviewee and his or her situation. Use phrases such as "Let me see if this is what you mean…" and "Tell me if I understand you correctly…" If the interviewee believes you really want to understand, then you don't have to reflect their feelings, thoughts, and behaviors perfectly.

At this point you also have to still your own thoughts; this is difficult to do because we are continually processing our own thoughts and reactions to what the client is doing or saying. The cognitive processing that occurs as we listen includes evaluative judgments. These can interfere with how we hear the client. For example, the client says, "My mother is so difficult to talk to. And I usually end up yelling at her." Initial thoughts such as "you and your mother have a poor relationship" or "you have aggressive tendencies" may preclude your actually being with the client.

Helpers often use unique approaches to establish rapport. For example, Cristofor, a staff member in a residential facility for teen youth, met with a new resident, Suzanne, for the first time this afternoon. Cristofer will be one of the weekday, after-noon/evening staff working with Suzanne during recreation, homework, and group therapy. He began his initial interview with Suzanne by asking her to share with him the three most important things she wanted him to know about her. This is his way of beginning the relationship on her terms. Christofer believes this is important because many of the residents come to the facility under court orders.

A case manager for persistently mentally ill clients has the initial interviews in her office. Many of her clients have been receiving some type of human services since they were adolescents; they have had interactions with numerous helping pro-fessionals. She has a large 5- by 6-foot storyboard that her clients use to map out the story of their lives. They configure their families and significant events, moving pieces around the board as they share some of their social history. Her undivided attention to them and their stories help her establish rapport in the initial setting.

What Determines if You have Successfully Established Rapport with the Interviewee?

When the interviewee believes that you "get it" and you have "heard" her view-point. Many interviewees will grant you credit for trying. Be cautious though. Sometimes you will encounter a client whose world or behavior is so foreign to you that you simply are unable to understand her perspective. Even so, you can lis-ten carefully, clarify, and hear the client's words.

> I ended up getting a really wonderful worker. When I went in to see Sheila, she could... I don't know, we just clicked and it was almost like she said, "You need to cry." And I said, "Yeah." Because at that point I was living in a women's shelter in another county. So I was still walking on eggshells. I did not know where to go or what to do. I had no self-esteem, no self-respect. Sheila gave me the help, verbally and emotionally, that I needed. The words that I needed to hear for so long, the words of praise: "Hey, you have finally gotten out of it."

Sheila was there with caring and compassion.

 EXERCISE 3.1: **Establishing Rapport**

Video Clip A illustrates an initial meeting between an interviewer and an inter-viewee. Watch and listen carefully. Then answer the following questions:

1. *Describe what happens.*
2. *Is rapport established? Why or why not?*
3. *What part does culture play in this interaction?*
4. *Did you detect any difficulties? What were they?*

What About That Client Whose World or Behavior Is so Different from Mine?

Undoubtedly you will encounter a person who is very different from you. The dif-ference may be religious, racial, or ethnic. Dress, customs, behavior, language, and values are other common areas of difference. How do you establish rapport with someone different?

In the helping professions, an important distinction exists between acceptance and approval. Acceptance is the ability of the helper to be receptive to the client regardless of the way a client is dressed or what the client may have done. An interviewer acts on this value of acceptance by maintaining an attitude of goodwill and refraining from making judgments. This interviewer behavior includes appreciating the client's culture and family background. Approval implies that you are in agreement with another person and that you will speak or think favorably about the person, the behavior, or both (Woodside & McClam, 2009). As you think about Marianne's home visit, how do you think she felt when she walked into a home with a dirt floor? Because I know her, I would suspect she was surprised but accepting of the condition of the family's home. To disparage their home or to communicate in any way that it was inadequate would have certainly established a different tone to the meeting.

How Do I Get Another Person to Trust Me?

When trust is established within an interview, the interviewee then has confidence in the interviewer's ability to provide help. The interviewer wants to establish trust to promote a positive interview environment, encourage the interviewee to self-disclose information that enhances success of the interview and other phases of the helping process, and engage the interviewee in the helping process (Tang & Russ, 2007).

An interviewer works to establish trust in several ways. One way is to act with honesty and integrity. Using language that the interviewees can understand, interviewers provide information about the purpose of the interview and their own role in the interview. They share a professional disclosure statement indicating their training, experience, and skills and the limits of the services they will provide. In addition, interviewers describe agency policies related to confidentiality and disclosure of information revealed during the interview process.

Within the interview process, interviewers indicate that they are engaged in two-way communication with the interviewee. Not only do they indicate they have heard the interviewee and are trying to understand the interviewee's world, but they also demonstrate a willingness to be influenced by them (Schoenholtz, 2000). For instance, interviewers might say, "I have never thought about the situation you describe in that way … it changes how I think…" or "I can see that our interview process does not completely work for your situation. Let me see if I can make some changes to our way of collecting information."

Another way to build trust is to keep your word. This means you need to be clear about the time you will be able to give, the services that are available, and the changes that may occur during the initial interview and beyond. Sometimes in the moment, interviewers want to assure those coming for help that assistance is "on the way" and that interviewee's lives can be transformed. Be positive, but realistic!

Are There Things I Shouldn't Say?

During the initial interview session, the interviewer welcomes the interviewee and makes an initial connection. Any behaviors that signal to the interviewee that you are disinterested, have preconceived opinions about her situation, or are disrespectful will hinder the interview. For example, if you frown at the interviewee or are tense or distant, she may infer that you do not want to provide help. If you make pronouncements about the interviewee's race, gender, socioeconomic status, family status, or other characteristics, you convey a stereotypic attitude. For example, statements such as "I know that this is a common problem for African Americans." "Your status as an immigrant must make you feel alienated." or "Being a single mother has few rewards." signal your biases and may stifle communication and prevent the development of trust. One interviewee shares her interview experience below. Note her reactions and how she perceived the intentions of her interviewer.

> *My second counselor treated me like dirt. "Here you are getting benefits and you are not doing anything to help yourself. You are not even trying to find a job." It was like I was a nobody. The initial interview was about getting me processed in the computer. And at that point, you become a number. And that is all she had time for.*

What Is the Best Way to Help the Interviewee Feel Secure in an Interviewing Situation?

Asking interviewees to tell their own stories in an informal way is one way to provide support to them during the interview. Leads such as "Can you tell me more about that?" "Describe a time when you had that experience?" and "Tell me your favorite story about ____?" provide openings for the interviewee to talk. Stories provide a way for them to make sense of their world and their challenges and share those thoughts and feelings with you. Listening to the stories signals to the interviewee that you believe that he is important and that his experiences have value. Many times if interviewees are willing to share their stories, it means they feel comfortable with you and trust you with information about themselves and their lives.

So How Do You Build a Relationship with a Person Who Comes Unwillingly; for Example, the Person Who Is Referred but Doesn't Want Help?

A person who comes to the interview against her will is often reluctant. Generally, the reluctance occurs at the beginning of a relationship and describes the person who, if given a choice, would prefer not to be present or talk about his or her self. This individual misses or is late to the first meeting or is uncommunicative once there. Other behaviors that might indicate reluctance are the denial that help is needed ("Mrs. Smith just doesn't like me."), expressions of grandiosity ("I'm sure you can make her come back to life."), excessive agreeableness ("Whatever you suggest is okay with me. I'll do whatever you say."), and insignificant information ("Yesterday I went to the movies."). These behaviors may arise from negative attitudes about helping, beliefs that getting help means weakness or failure, feelings of anger or embarrassment, and unfamiliarity with the agency or the helping process.

It is important to recognize reluctance and its causes and to develop strategies to use with the reluctant individual. Ignoring reluctance can negatively affect the initial meeting and impair the development of a relationship. For example, as reluctance increases, client dissatisfaction increases, and little progress occurs. Failure to recognize and deal with reluctance may even bring an end to the relationship, so don't ignore it!

How Do I Help a Person Like This Open Up?

There are several strategies you can use with reluctant individuals. First, recognize the person's feelings. This may mean acknowledging that a third party, perhaps the school, the court system, or the place of employment, referred them or required their presence. If this is the case, then the agency's purpose may be a mystery to them. A second strategy is to explain the process by discussing expectations, confidentiality, the interviewer's role, the individual's role, why the person is there, and what is going to happen. Displaying empathy and support to build trust is another strategy. When you show concern for and interest in another's perceptions about what is happening, you communicate support for that person. Patience and genuineness will also assist you in establishing the relationship. Otherwise, you might find that you are doing all the work in the interview. Once you acknowledge reluctance and begin to build a relationship, you are both ready to complete the interview and explore the presenting problem (Woodside & McClam, 2009).

EXERCISE 3.2: **Approaching the Reluctant Interviewee**

Watch Video Clip B. The interviewer begins an interview with a reluctant interviewee. Answer the following questions:

1. *How do you know this interviewee is reluctant?*
2. *Describe how the interviewer approaches the interviewee.*
3. *What strategies seem to be successful?*

What Kind of Nonverbal Behavior Will Help Me Establish a Relationship During This Initial Encounter?

Nonverbal communication is an important way that people communicate with each other. Experts believe that at least two thirds of the meaning of our communications is transmitted nonverbally, so it's a powerful mode of communication. Any conflict between verbal and nonverbal communication results in a person believing the stronger nonverbal message. Much of our nonverbal communication transpires through facial expressions, gesticulations, body language, tone of voice, silence, physical space, and touch; more often than not, it occurs spontaneously and without intent (Miller, 2005).

Interviewers can use nonverbal communication to make connections with interviewees, especially when used deliberately to convey availability to help. There are five types of nonverbal behaviors to assist interviewers in making a connection: eye contact, body position, silence, tone of voice, and facial expressions.

When relating to individuals in a Western culture, the helper maintains good *eye contact* to communicate "I am here with you" and "I am listening to what you have to say." This does not mean staring at the client without breaking away;

rather, eye contact is used as an encourager. It is relaxed, and breaks away naturally when the listener is nodding the head or thinking about what the helpee is saying. In other cultures or in different contexts, direct eye contact may have alternate meanings, making it critical that the interviewer be aware of the meaning of eye contact to people with different cultural backgrounds. Whether it is limited eye contact or a sustained focus, looking at the interviewee gives the interviewer important insights into what the interviewee is saying by attending to the interviewee's nonverbal language. Eye contact, then, has important implications for both the interviewer and the interviewee (Miller, 2005).

Body position that encourages the interviewee is most often "relaxed tension." This is the posture of involvement. It means the interviewer faces the other person squarely, leaning forward slightly. A balance is achieved when the posture communicates, "you have my attention" and "I am interested in what you have to say." If the interviewer appears too tense or is leaning too far forward, the interviewee may sense a demand to talk (Kader & Yawkey, 2002).

Many beginning professionals are uncomfortable with *silence* during any client contact, but especially during this initial meeting. Actually, silence is an important nonverbal communication tool. It provides time for interviewees to process what is being said, to think about what they are feeling, and to decide what they wish to say. Silence will be addressed more fully in Chapter 4. In this initial meeting, however, it is important that the interviewer doesn't allow silence to become a deterrent to the dialogue that is occurring. Pay attention to the pace and flow of communication; allow silence to occur, and move the interview along.

Interviewers use *tone of voice* to communicate a sense of calm, stability, and order to the interviewee. Many people come for help in a troubled emotional state or from chaotic lives; sensing calm from the interviewer provides reassurance and encourages confidence in the work they will do together to resolve the problem or situation. Other times interviewers may mirror or match the interviewee's emotional state or emphasize certain words in order to assure her a depth of understanding and empathy (Kader & Yawkey, 2002).

Facial expressions are an interesting component of nonverbal communication. We believe that few of us have a good understanding of our own facial expressions and what they mean to others. Often we are simply unaware of our own facial expressions. This is why many academic programs that prepare helping professionals include videotaping as a way to develop skills. Often, too, we don't understand the other person's frame of reference for decoding our expressions. This same phenomenon occurs in an initial meeting. Neither participant may know enough about the other to decode facial expressions. A later chapter will provide the knowledge that will help develop your skills in this area.

How Do You Use Verbal Communication to Build the Relationship?

It's important to remember that if you are talking, then you aren't listening! Many beginning professionals have a tendency to talk too much. Whether this occurs out of their own anxiety or simply that they want to talk, it results in an interviewer who is doing all the work and who may find, at the conclusion of the meeting, that necessary information has not been gathered. In Chapter 4, the use of questions and other responses is introduced, and you will have opportunities to further develop these skills.

You have read about introductions and door openers. These are ways to use verbal communication to build the relationship. Openers that place the decision about the topic in the hands of the interviewee are phrases such as "So what's on your mind today?" "Tell me about you." or "You seem upset. Let's talk about what's going on." These phrases also demonstrate to the interviewee that you are paying attention. In addition to openers that signal you want to hear from the interviewee, you can also use what we call verbal following: yes, hmm, I see, and right. These short encouraging responses tell the interviewee that you are listening to them. They may be invaluable in connecting with the interviewee.

It Seems Like I Am Getting a Sense of Working Together with the Client. What Can I Call This?

Some interviewers are comfortable thinking of the relationships as a partnership, others think of this as developing a working alliance (Flicker et al., 2008; Kramer et al., 2008). Within the frame of helping, several characteristics of a partnership or alliance make this term appropriate to use. First, a partnership denotes an agreement between parties to work together on agreed upon goals. In an initial interview, this can occur, especially if the interviewee is clear about the purpose of the interview and the interviewer has clearly explained his role and responsibility. A second characteristic of a partnership is mutual cooperation. The initial spirit of cooperation is offered and freely given by the interviewer.

A third characteristic of partnership or alliance is mutual responsibility. During the initial interview, the responsibility of the interviewer is to establish a positive climate, explain the nature and purpose of the interview, and describe the process of helping and the ethical bounds of the helping relationship. Responsibilities for the interviewee might include identifying strengths, needs, and information about context. In reality, the interviewee may accept or reject these responsibilities. Again, responsibility for establishing a positive climate that increases the likelihood that the interviewee accepts these responsibilities lies with the interviewer.

COMMON BUT DIFFICULT: **Dealing with the Client's Strong Emotions**

You are interviewing Christopher who begins to cry as he tells you that his partner, who has AIDS, has been hospitalized. It is doubtful that he will come home this time.

Often interviewees will share a situation that involves strong emotions. Intense feelings of happiness, anger, fear, or sadness may cause tears that trigger different responses in interviewers. One is dismay or discomfort with crying and puzzlement about how to respond. Another response is denial and reassurance: "Don't cry. Everything will be okay." Unfortunately, you don't really know that everything will be okay. In Christopher's situation, he probably believes right now that nothing will ever be okay again.

Crying is one healthy way to release intense emotions. Talking with someone who is genuine, empathetic, and concerned may prompt the release. It's a good idea to keep tissues on your desk for these situations. We can also acknowledge tearful emotions by responding appropriately. For example, you might say to Christopher, "Oh, Christopher. This is such sad news. I can see how upset you are." Or you may say nothing; instead a touch on the hand or arm may say all that needs to be said at that moment.

CULTURAL CONTEXT: **Nonverbal Behavior in a Cultural Context**

Earlier in this chapter you read about the kind of nonverbal behavior that helps the interviewer establish a relationship with the interviewee during the initial encounter. Thinking about nonverbal communication within a diverse cultural context requires knowledge of the culture from which the client comes and attention to client responses during the initial interview. Our discussion centers on interacting with members of groups with less power, privilege, or economic, political, educational, or social status.

Interviewees from minority groups may come to the interview with a sense of cultural mistrust or "healthy distrust of potentially threatening environments and systems" (Atkinson, p. 94); this is especially true of African Americans, who often see education and other systems as White institutions (Atkinson, 2004). Results of cultural mistrust are resistance to asking for help, less satisfaction with services, reduced self-disclosure, and negative evaluations of helpers.

If you, as the interviewer, are a member of the dominant culture and wish to convey respect to the interviewee, then attending to the ways in which you communicate is important. You might indicate respect with a slight bow when the individual comes in, allow her to choose where she sits, have chairs that are of equal height, sit without a table or around a table rather than across a desk, be on time, and be prepared. Other behaviors are making eye contact, leaning forward slightly, using minimal encouragers, and providing space in the conversation for the interviewee to talk without interruption.

ETHICAL CONSIDERATIONS: **Value Conflicts**

An ethical situation may arise when the interviewer discovers during rapport and relationship building that a value conflict exists between herself and the interviewee. Sensitive areas of potential conflict include, but are not limited to, moral, religious, or political values (Tjeltveit, 1986); different worldviews and philosophies; and/or unacceptable or egregious behavior as defined by the interviewer. When interviewers encounter strong resistance or reluctance to developing rapport and establishing a relationship with an interviewee, the question arises about referral.

We believe that the decision to refer should be made based upon what is best for the interviewee. Value conflicts and discomfort, at times, are a matter of degree. The interviewer must gauge her competence in the area of conflict, discomfort with the interviewee, and the ability to be objective when viewing the interviewee and his responses (Corey, Corey, & Callanan, 2007) to determine if the interview can proceed with benefit to the interviewee. Complicating factors concerning referral are a feeling of rejection by the interviewer, the perception of being given the run-around, or a limited number of professionals to whom the interviewee can be referred.

One example that demonstrates the complexities of value conflicts is an interview with a woman who had previously belonged to a cult that took her baby from her at birth. She escaped the cult without bringing her baby along. The cult had ritual sacrifice of children; the interviewee showed no remorse or concern for her child. As an interviewer, what might be your reactions? This is only one example of the many occasions when value conflicts between interviewer and interviewee may occur.

Personal Reflections

1. *Have you ever felt you had rapport with another person? How do you think this happened? Describe how it felt to you.*
2. *What does trust mean to you? Who do you trust? Why do you trust? Is it possible to trust someone else totally?*

References

American Heritage Dictionary. (2000). *Rapport*. Boston: Houghton Mifflin.

Atkinson, D. R. (2004). *Counseling American minorities* (6th ed.). Pacific Grove, CA: Brooks/Cole.

Brill, N., & Levine, J. (2004). *Working with people: The helping process*. Boston: Allyn & Bacon.

Cochran, J. L., & Cochran, N. (2006). *The heart of counseling: A guide to developing therapeutic relationships*. Pacific Grove, CA: Brooks/Cole.

Corey, G., Corey, M. S., & Callanan, P. (2007). *Issues and ethics in the helping professions* (7th ed.). Pacific Grove, CA: Brooks/Cole.

Flicker, S. M., Turner, C. W., Waldron, H. B., Brody, J. L., & Ozechowski, T. J. (2008). *Journal of Family Psychology, 22*(1), 167–170.

Kader, S. A., & Yawkey, T. D. (2002). Problems and recommendations: Enhancing communication with culturally and linguistically diverse students. *Reading Improvement, 3*(1), 43–51.

Kramer, U., de Roten, Y., Beretta, V., Michel, L., & Despland, J. (2008). Patient's and therapist's views of early alliance building in dynamic psychotherapy: Patterns and relation to outcome. *Journal of Counseling Psychology, 55*(1), 89–95.

Miller, P. (2005). Body language in the classroom. *Techniques, 80*(8), 28–30.

Okun, B. F., & Kantrowitz, R. E. (2008). *Effective helping: Interviewing and counseling techniques* (7th ed.). Pacific Grove, CA: Brooks/Cole.

Schoenholtz, S. W. (2000). Teaching as treatment: A holistic approach for adolescents with psychological problems. *Preventing School Failure, 45*(1), 25–30.

Tang, M., & Russ, K. (2007). Understanding and facilitating career development of people of Appalachian culture: An integrated approach. *The Career Development Quarterly, 56*(1), 34–46.

Tjeltveit, A. C. (1986). The ethics of value conversion in psychotherapy: Appropriate and inappropriate therapist influence on client values. *Clinical Psychology Review, 6*, 515–537.

Woodside, M., & McClam, T. (2009). *Introduction to Human Services* (6th ed.). Pacific Grove, CA: Brooks/Cole.

4 | # Skills and Strategies

One of the biggest challenges of working with adolescents is making sure that referrals happen. At our agency once we identified a client need that we couldn't meet, we had a list of state-approved professionals who would see our clients at reduced rates mutually agreed upon by both parties. The list included physicians, psychologists, physical therapists, and specialists like neurologists and psychiatrists. The problem was clients not showing up for their appointments. Often the referral was made, the client was informed, but the client didn't follow through and show up. I was never sure exactly what went wrong, but I sometimes wondered if the problem was ours rather than the client's. Did we explain enough? Did we offer enough support? Was there transportation? Did the client just blow us off? I've often wondered about our role in referring clients. TM

It is an accepted reality in the helping professions that often one person, one agency, or both cannot meet all client needs. One goal of the initial interview is problem identification. A second goal is developing a strategy to address the problem or problems. These are critical tasks and their successful accomplishment influences the rest of the helping process—even the effectiveness of the referral process that Tricia wondered about. We've discussed communication skills in the previous chapters, for example, making introductions, building rapport, and establishing trust. This chapter will review some specific skills that promote problem identification and referral.

What Are Those Skills that Will Help Me with Problem Identification?

One key interviewing skill is listening. It is an intensive activity that takes concentration and time. Interviewers listen to what is said both verbally and nonverbally; but they also listen for what is not said. Good interviewers engage in listening behaviors, sometimes called *attending behavior* or *responsive listening*. These behaviors let interviewees know that they are being heard. Interviewers maintain eye contact if culturally appropriate, use verbal following (yes, I see, hmmm), and have open body language. Egan (2007) suggests a framework of skills that will help you visibly tune in to interviewees. We have used them over the years and found that they work well. They can be summarized with the acronym SOLER: Face the other person Squarely; adopt an Open posture; Lean forward slightly; maintain Eye contact; and have a Relaxed posture.

Another key skill is exploring. Sometimes identifying the problem is like peeling away the layers of an onion. As the relationship moves to a greater depth of involvement, trust, empathy, and genuineness are experienced, creating an intimacy that encourages the freedom to express whatever thoughts or feelings the interviewee is experiencing. This climate then allows for the exploration of the presenting problem and the identification of any underlying problems. The following excerpts start with the same presenting problem but the interaction moves in two different directions.

EXCERPT 1

INTERVIEWEE: I decided to come in today although I have a raging headache. In fact, I've had a headache each day this week.

INTERVIEWER: I've got some aspirin right here. Now, how are you?

EXCERPT 2

INTERVIEWEE: I decided to come in today although I have a raging headache. In fact, I've had a headache each day this week.

INTERVIEWER: I'm glad to see you but sorry you aren't feeling well. You must be a bit frustrated that they keep recurring. What's going on?

INTERVIEWEE: I get these headaches first thing in the morning before I've even started my day. Then my neck tightens up and I haven't even had breakfast at that point.

INTERVIEWER: And I suspect you feel like your day is already off to a bad start.

INTERVIEWEE: I do. And before I know it, I'm snapping at my husband and the kids so then everyone's day is off to a bad start.

INTERVIEWER: Sounds like you are stressed. Let's talk about this.

Clarifying is another skill that is useful in problem identification. It helps interviewers and interviewees sort out what is going on and focus on understanding the basic message. What is the interviewee really saying? What does the interviewee really mean? Often the interviewer may be confused or unsure about what was said (or not said) because the situation, behaviors, or feelings being shared are complex or conflicting. It is important to stop at that point to clarify rather than to assume you know what's going on.

INTERVIEWER: This is a complex situation and I want to be sure I understand what the issues are. Could you go over that again?

INTERVIEWER: I'm confused about how you really feel about what happened. Were you embarrassed that Joan was there and saw what happened?

INTERVIEWER: Sounds to me like you wanted her to say yes.

Often there are underlying issues that cause a person to seek help. Taking the time to thoroughly listen, explore, and clarify pays big dividends as the interview develops. Look again at Excerpts 1 and 2. The interviewer in Excerpt 1 has missed what the interviewee is trying to say by moving too quickly, and then compounds the problem by asking how the interviewee is immediately after the interviewee has shared that very information. This interviewer will spend even more time trying to return to what the interviewee said once it is clear the problem hasn't really been identified or the interviewee will leave with the aspirin but no help with what is really going on.

Is it Okay to Ask Questions?

Questions are a natural way of communicating and an important technique to elicit information. This is after all the purpose of an intake interview. There are some advantages to questioning. One is that questions save time. You know exactly what information you need, and questions are a direct way to get it.

INTERVIEWER: When is your birthday?

INTERVIEWEE: August 1.

Second, a question can focus one's attention in a particular direction. Often an interviewer may want to clarity an inaccuracy or an inconsistency:

INTERVIEWER: You said you hit people when you are angry and that's what you did to Sam. But you don't hit your mother. How is the fight with Sam different from the fight with your mother?

A question can also move the dialogue from general to specific or specific to general.

INTERVIEWER: You've talked about how angry you get at times at school. What happens when you get angry with a classmate? (general to specific)

INTERVIEWER: You described what happened when you got angry with your friend Sam. Could you describe how you feel when you're think you're out of control? (specific to general)

It is true though that questioning can be a pitfall for a beginning helper (Okun & Kantrowitz, 2008). One question may lead to another and another and another until the interviewee may feel like it's an interrogation. Resulting feelings may include defensiveness, hostility, or resentment. It's a difficult pattern to break once it begins. Another problem is the tendency to ask questions when there is silence. A period of silence may be uncomfortable and a question can fill it. Questioning can also limit self-exploration and place the individual in a dependent role where his or her only responsibility is to respond.

Questioning is an art and beginning helpers might want to avoid relying on questions. One way to do this is to reword your questions as a statement. Egan (2007) suggests that if you are going to question, then consider how the question will relate to and promote the rapport and the interview.

When Is a Question a Good Idea?

There are some times when a question is appropriate and will help you move the interview along. They include beginning an interview ("Could you tell me a little about yourself?"), to obtain specific information ("When have you felt the same way?"), to clarify ("What is different about the two situations you described?"), and to focus the client's attention ("Of the challenges you've talked about today, which one is the most immediate concern for you?").

Another time when questioning is helpful is the identification of interviewee strengths. Helping professions today have moved away from a deficit model to one that emphasizes client strengths. Have you encountered this problem in the past? What resources helped you solve it? What did you do to keep it from turning into a crisis? How have you solved problems in the past? What worked for you then? Answers to these questions will promote the identification of the positive characteristics, abilities, and experiences the interviewee has had. This information may be helpful in promoting growth and change as well as helping clients approach problem areas by using their past successes.

What Type of Questions Should I Ask?

We are familiar with the how, what, when, where, and why question sequence. These types of questions focus on the cognitive domain, that is, the facts rather than the emotions or feelings of the individual. Too many are problematic. We even recommend removing "why" from your questioning repertoire. "Why" questions breed interviewee defensiveness and ask for a justification for a behavior or belief. When interviewers use a "why" question, they often create more barriers to establishing a relationship than need be.

The questions used in intake interviews can be categorized as either open or closed inquiries. Determining which one to use depends on the interviewer's intent. If specific information is desired, closed questions are appropriate:

INTERVIEWER: Are you married?

INTERVIEWER: What grade did you complete in school?

INTERVIEWER: Do you use drugs?

These questions elicit facts. These are similar to the questions you might find on a form that requires a yes, no, or simple factual statement. One caution is to be careful not to use a series of closed questions that cause the interviewee to feel interrogated rather than helped. Often a series of closed questions occurs when there is a form to be completed during the interview. It is difficult *not* to let the form take precedence over the establishment of a relationship.

If the interviewer wants the client to talk about a particular topic or elaborate on a subject that has been introduced, open inquiries are preferred:

INTERVIEWER: How would you describe your marriage?

INTERVIEWER: What was school like for you?

INTERVIEWER: What is your experience with drugs?

These questions are broader, allowing the expression of thoughts, feelings, and ideas. They require a more extensive response than yes or no. Open inquiries also contribute to building rapport and identifying problems.

INTERVIEWEE: My child is having trouble in school. His grades are falling, and it seems like he's always fighting about something.

INTERVIEWER 1:   How old is he? (OR what grade is he in? Does he have any friends? What classes is he in? Who is he fighting with? How long has this been going on?)

INTERVIEWER 2:   Could you tell me more about what's going on?

Interviewer 1 responds with a closed question that requires a simple answer; the follow up response will likely be another closed inquiry. Interviewer 2 asks the interviewee to elaborate on what he thinks is going on with the child. The interviewee is free to introduce topics and share information. The interviewer is free to encourage exploration and clarification of this parent's concerns.

How Do I Ask A Question?

Wording is often less important than manner and tone of voice. Strike the right tone by being genuine and interested; pace is a delicate balance between too fast (don't get all the information) and too slow (disinterest). Remember that questioning is an art and until you develop your skills in this area, any question can be re-phrased as a statement. And you may find that once you develop this skill, asking a question actually becomes challenging.

What Are Some Other Ways to Elicit Information During the Interview?

Most interviews are an exercise in data gathering, and we know that question after question may impede the sharing of information. You've read about some of the negative outcomes of too many questions. There are other techniques for gathering data. "Tell me more about...," "Describe how or when....," Let's explore....," and "I'm wondering if...." are all nonthreatening ways of acquiring information with few negative consequences. Prefacing questions with some of these phrases helps defuse defensiveness, reluctance, and the feel of interrogation.

EXERCISE 4.1: **Problem Exploration**

Watch Video Clip A to see a skilled interviewer demonstrate problem exploration with an interviewee. After watching the video, answer the following questions:

1. *Identify the behaviors that communicate the interviewer is listening.*
2. *What techniques does the interviewer use to explore the problem?*
3. *Describe how the interviewer uses questions.*
4. *In what other ways was information elicited?*

What Is Referral?

Referral is the connection of an interviewee with a service provider other than you. You may refer "in house" (within the agency) or to an outside professional. For example, the agency that employed Tricia also employed a psychologist to provide basic psychological testing. As you might imagine, this professional was more often than not overwhelmed with "in house" referrals—particularly from those counselors who worked with troubled adolescents or adults with mental or emotional problems. Sometimes, it was necessary for counselors to refer their clients to a psychologist who had a contract with the agency to conduct psychological evaluations. In both cases, the individual was linked to other professionals for necessary services.

I've Identified a Client Need that I Can't Meet. Is This When I Refer?

You have already determined that there is no match between the client need and the services you provide. Do you end the relationship or refer? Helpers have a responsibility to locate resources in the community, arrange for the interviewee to make use of them, and support the interviewee in using them. So this is a time to refer.

There are other times to refer. The individual you are interviewing may request a referral. This person is aware of a service and believes that a referral from you will give him more credibility with the new service or that you can speed up the response. The interviewee may feel that the two of you cannot work together for some reason and requests a different helper. Another reason for referral is that the individual may be in serious psychological distress—suicidal or violent—and require intensive therapy or crisis intervention that the interviewer cannot provide. Finally, you may decide to refer because of lack of progress with a particular interviewee and perhaps a sense that a match with another professional might be more productive.

There is one caution here. If you refer all your interviewees (unless of course, that is your role as an intake interviewer), then you may find yourself out of a job! The inclination to overuse referral may be a sign that you would benefit from professional development, additional supervision, or training in identified areas of client needs.

Making a Referral Doesn't Mean I've Failed, Does It?

Referral is not failure. In fact, there are several advantages to referral. One is access to an array of services for the interviewee. Particularly in times of economic downturns and budget constraints, agencies are forced to limit their services and instead, choose to offer fewer but more focused services. It is realistic to think that you might be employed by such an agency. Second, as helping professionals, we are ethically limited to providing only those services that we are competent to provide. In fact, you should carefully read your ethical code so that you are clear about your boundaries. Service coordination among helping professionals and agencies enables us to link our interviewees with those who have the expertise in needed areas. Finally, the people we work with have a right to receive the services they need without getting the runaround or encountering frustrating confusion among helpers.

I've Decided to Refer. What's the Key to a Successful Referral?

There are a number of things you can do to increase the likelihood of a successful referral. One is to explain to the client why the referral is taking place. This is also true for client-initiated referrals. The individual needs to understand that he is not being passed off to another helping professional. Rather, this is a connection that is in the best interest of the client. An explanation might be similar to the following:

> *Roger, I know you want to improve your parenting skills, and I think doing so will help not only your relationship with your children but also with your wife. There is a men's parenting group that meets every two weeks at the community center on Elm Street. It's free and the group commits to confidentiality. You might find it helpful to talk with other parents who are having similar experiences. In fact, I know the group leader, and if you'd like, I can give him a call to find out exactly when the meeting is and if there's anything you need to do.*

This interviewer has also described what she knows about the group. The more information you can provide about the agency and/or the services, the more you demystify the referral and perhaps relieve some anxiety. In this case, the interviewer knows the group leader, which is helpful for the interviewee and contributes to a feeling of connectedness. If you don't know anyone, then you may want to call the agency yourself for the name and number of a contact person.

INTERVIEWER: I talked with an individual this morning who is concerned about getting tested for AIDS. Is there someone I could talk with about this service? (Notice no identifying information is provided at this point.)

RECEPTIONIST: Jan Johnson is the person you need to speak with. Her number is 333-4567. Let me transfer you to her office.

INTERVIEWER: Jan, this is Alice at Haven Center. I talked with someone this morning who wants to be tested for AIDS. Could you tell me how to go about this? Also, how much does it cost, how long does it take, and when and how are the results communicated?

Other information you may want are directions, hours, and the person your interviewee should call. Will she be seeing Jan or someone else? The more complete the information the more likely the interviewee will follow up on the referral.

Another tip for a successful referral is transmitting information to both the client and the agency or service. The new service may need background information on the client. Has the client signed a release? The client also needs to know as much as possible about the referral. And you will need to determine if the client is capable and willing to follow through or if you should call to make the appointment for the client.

EXERCISE 4.2: **Referral**

Watch Video Clip B to see a referral in action. At this point in the interview, the interviewer has determined that a referral is necessary and discusses it with the interviewee. After watching this clip, answer the following questions:

1. *How does the interviewer introduce the idea of a referral to the interviewee?*
2. *What skills does the interviewer use?*
3. *How does the interviewee react?*

How Will I Know Where to Refer?

Knowledge of community resources is a valuable resource when making referrals. Many communities publish directories of agencies and services. For example, the entry below comes from one such director and illustrates the kinds of information you might expect from an entry.

RESPITE

Short-term relief for the family caregiver by providing a trained substitute caregiver, either in or outside the home. **Adult day services** (pages 41-44) and **homemakers & in-home services** (pages 87-88) also provide care for an older person and time off for a caretaker. Some **nursing homes** (pages 133-136), **retirement centers, assisted care living facilities, and residential homes for the aged** (pages 96-102) offer weekend and vacation respite, space permitting, but be aware that space, especially in nursing homes, is often not available. Private pay; some long-term-care insurance policies cover respite.

SAMARITAN PLACE.......................................555-5555

Joint partnership between Catholic Charities and St. Mary's Health System
900 E. Oak Hill Avenue, 23456-7890

24-hour supervised facility for individuals 55 and over who need temporary supportive care. Private pay.

Both formal and informal networks are useful in determining what services are currently available in a community. Tapping into these resources is relatively easy. Talk with co-workers, attend professional meetings, and participate in professional development opportunities. These are excellent ways to network. In one community we know, there is a monthly luncheon meeting open to anyone who is a helping professional. There is lunch, of course, but also a brief presentation about a service, an agency, or even an identified need. One task of the interviewer who makes referrals is to build a personal card file or electronic database of other professionals and agencies, their locations, hours of operations, and services.

What Communication Skills Are Going to Be Most Useful?

You've already read about the role communication skills play in establishing a relationship with a client. It is no different between professionals. Many helping professionals work with their counterparts in other agencies and offices. They receive referrals from them as well as make referrals. One important aspect of successful service coordination depends on the interviewer's relationship with other professionals. The following suggestions show ways to enhance communication with other helpers.

First, avoid stereotyping other professionals. You may have encountered one counselor or teacher who was rude, but it is unreasonable to think that all counselors or teachers will be that way. Second, recognize what you don't know or understand, and ask for clarification or a definition of terminology. It is better to ask than to pretend that you know. At the same time, you may be asked questions that you can't answer. Don't bluff. Feel free to say "I don't know." or "I'll find out." Finally, be aware that other professionals may well have different styles of communication. Styles that you may encounter in your work with other professionals are clinical (medicine), legal (of equal adversaries), political (of unequal adversaries), and pedagogical (teacher–student). Specific behaviors that help you achieve more effective communication include listening, avoiding technical jargon, identifying mutual concerns and goals, and focusing on problem solving.

Do I Need to Evaluate Referrals?

Evaluating referrals is strongly recommended. You don't want to continually refer people to an agency that doesn't deliver or treats them rudely or never sends reports. Systematic follow-up can be time consuming, but it will provide you with needed information. You can call to find out if the client kept the appointment, and you can ask the client for an evaluation. This information will help you build your own referral system, is good feedback for both the client and others, and increases the client's faith in the system.

COMMON BUT DIFFICULT: **The Client Who Does Not Want to Be Referred**

The interviewee has invested time and energy making an appointment for the interview, comes to see you at an appointed time, and talks with you about the issues and challenges that he or she faces. You, as the interviewer, have worked to establish a climate of trust, build rapport, and determine client strengths and problems. As stated earlier, there are times when the services you offer and the knowledge and skills you have do not meet the client needs, and you believe you need to refer. The client resists the idea of referral and counters, "Why don't I just continue to work with you!" "I don't want to go anywhere else for help." "I don't have the courage to talk about these problems with anyone else." It is a natural reaction to feel committed to the client and to feel the tug that loyalty to the client brings. And you don't want your client to give up on getting help.

You have already learned about making a successful referral. What additional measures can you take in the case of the reluctant referral? Medical professionals provide some excellent advice for helping professionals. Research indicates that patients do not follow up with referrals because the wait is too long for the appointment and/or they do not believe that the referral is necessary (Wu, Kao, & Chang, 1996). Sometimes clients do not want to make an appointment, stating they want to think about follow-up or indicating they don't have a calendar with them or cannot insure they will have transportation (Nutrition Works, 2007).

Making the successful referral means being clear why the referral is necessary and articulating clearly what the client gains by seeing another professional. Linking the referral to concrete gains improves client satisfaction with the services (Woodside & McClam, 2009). Using standard forms to make referrals provides consistency in the referral process. Remember to provide the interviewee with a copy of the referral form; this increases success of the referral (Burgess, 2002). These forms should include referring agency, phone, fax, and address, referral recipient contact information, details about future appointment, reason for referral, and date for follow-up between the two professional agencies. This way the client holds a copy of the referral and feels more empowered in the referral process.

CULTURAL CONTEXT: The Influence of Culture

An important professional challenge for today's helping professionals is preparation for the diversity that characterizes the United States, and consequently, the consumers of the helping professions. The racial and ethnic makeup of the United States has changed more rapidly since 1965 than during any other period in history (Surgeon General, 2005). There is movement of both immigrants and citizens within the United States. These demographic shifts have three important implications for the professional development of helping professionals.

First, it is critical that we recognize and acknowledge the influence of culture on those individuals or families who might seek help. How people from a given culture express and manifest symptoms, their coping styles, the family and community supports, and their willingness to seek help are all determined by culture. A more specific example is the syndromes that are identified as culture specific, culture bound, or cultural related. These terms describe clusters of symptoms more common in some cultures than others. Traditional Native clients may have Wacinko, the feelings of anger, withdrawal, or suicide that result from reactions to disappointment and interpersonal problems. Some Latinos, especially women from the Caribbean, have *ataque de nervios*, a condition that includes screaming uncontrollably, attacks of crying, trembling, and aggression. Without knowledge of these syndromes, an interviewer might make incorrect assumptions, resulting in inappropriate action.

A second implication of a diverse population is the difference between individualism and collectivism. Many of us accept and value the importance of the "self." This Western perspective deems independence desirable and the elimination of dependency a goal. To that end, we strive toward self-actualization and self-realization through self-help, self-awareness, self-disclosure, and so forth. In fact, this focus on the individual is a minority perspective in the world and often viewed by other cultures as exotic. In most cultures, like India and China, for example, it is normal and natural to consider the welfare of the family before any individual. Arranged marriages, a practice that intrigues many of us as so alien to the Western notion of romantic love, is a decision in some cultures that is uniquely important to the larger, extended family. It is to the benefit to the larger group, which takes precedence over the individual. In some cultures dependency is healthy and necessary. Daughters trust parents to arrange their marriages, and eldest sons are committed to assuming responsibility for elderly parents. Interviewers need to work comfortably with both individualistic and collectivistic perspectives.

Finally, if we accept these differences, then we must recognize that one source of strength for interviewees who are different culturally from the interviewer is the indigenous community. This may include religious, spiritual, and women's groups, or civic organizations. An astute interviewer will mobilize natural support systems as allies and include them as part of the helping process. This may mean referring the interviewee to a healer in his culture, using indigenous healing techniques, and/or consulting with traditional healers or religious and spiritual leaders and practitioners. Respect for the work of traditional healers is part of building the relationship with the interviewee. For example, traditional healers in Mozambique called *curandeiros* use herbs, potions, and cleansing rituals. While an interviewer may view these techniques as strange or silly and dismiss them, disparaging them in the interview will negatively affect the interaction between the interviewer and the interviewee.

The challenges to all helping professionals, but particularly those on the front line like interviewers, are to know your own culture and to learn about other cultures. This means reaching within your own self and reaching out to others. One way to accomplish this is to be aware of and develop a working relationship with community elders or religious leaders as partners in the helping process.

ETHICAL CONSIDERATIONS: **Respect for Interviewee Rights**

Respect for client or interviewee rights is a value that guides the work of helping professionals and directly applies to the initial interview. These rights are supported by professional codes of ethics, legal statues, and guidelines for delivering quality services. The rights extend from the time that the interviewee makes contact with the agency or social service system to beyond the termination of services. You, as an interviewer, have two responsibilities with regard to client rights: First, you need to know what rights clients have; second, you need to explain to interviewees what their rights are and incorporate these rights into the initial interview.

What are the ways you can introduce the rights of interviewees in the initial interview? Many agencies and professionals have handbooks, brochures, flyers, and letters that detail the rights that protect their clients. Providing this information to the interviewee at the initial interview establishes a tone of respect. What is critical in this communication is that rights are stated in language that interviewees understand and in concrete terms that reflect the actual realities of the interview and other services. For example, as discussed earlier, if the interviewer begins the initial contact with a discussion of informed consent, the interviewee receives the message, "Your rights are important to me." This means discussing informed consent in terms of client participation in treatment planning, service delivery, and refusal of services. Explaining agency policies helps interviewees know what to expect from the first interview. Professional disclosure provides interviewees with information about you, as the interviewer: your training, expertise, and responsibility during the initial interview.

Another key area related to client rights targets client information and release of information. A release of information requires a client signature, designates the recipient of client information, and indicates a period of time for which the release of information is valid.

The Health Insurance Portability and Accountability Act (HIPAA) regulations discussed earlier, outline measures that must be taken to meet client rights related to confidentiality. These include

- providing a statement that describes the agency policy with regard to client information, records, and safeguards to confidentiality.
- asking interviewees to sign a statement that they have been informed of the policies. (The signature verifies they have been provided this information.)
- assigning an individual responsible for oversight of the HIPAA regulations and policies.
- providing onsite safeguards for security of information.

Because agencies retain files after services have been terminated, commitment to safeguarding information extends beyond the initial interview and the end of service delivery.

There is a multicultural dimension to client rights with respect to power and authority. Many groups, such as women and African Americans, to name two, come to helping situations believing that they have little power (Sue & Sue, 2007). Articulating client rights at the beginning of the initial interview helps establish a tone of respect that facilitates the interview process.

Personal Reflections

1. *Your best friend has been sick for two weeks and has missed classes during that time. You know that he doesn't know what to do about the missed classes, papers, and tests. How can you use the referral process to help your friend?*

2. *Think about a time when you were referred to a professional for some reason. How did the referral occur? Was it effective?*

3. *Think about your last visit to a professional's office. What specific evidence did you see that illustrated a consideration of your rights as a client or patient?*

References

Burgess, I. (2002). Attention deficit hyperactivity disorder development of a multi-professional integrated care pathway. *Psychiatric Bulletin, 26,* 148–151.

Egan, G. (2007). *The skilled helper: A problem-management and opportunity development approach to helping.* Pacific Grove, CA: Brooks/Cole.

Nutrition Works. (2007). *Minimizing referral failure.* Retrieved August 20, 2008, from http://www .nutritionworks.us/silver/images/PDF/newsletter_ 0207.pdf

Okun, B. F., & Kantrowitz, R. E. (2008). *Effective helping: Interviewing and counseling techniques.* Belmont, CA: Thomson.

Riche, M. F. (2000, June). America's diversity and growth: Signposts for the 21st century. *Population Bulletin, 55*(2), 3–43.

Sue, D. W., & Sue, D. (2007). *Counseling the culturally diverse: Theory and practice* (5th ed.). New York: John Wiley.

Surgeon General's Report. (2005). *Mental Health: Culture, race, ethnicity: Supplement to Mental Health: A report of the surgeon general.* U.S. Department of Health and Human Services, Office of the Surgeon General. Retrieved August 9, 2005, from http://www.mentalhealth.org/cre

Woodside, M., & McClam, T. (2009). *An introduction to human services* (6th ed.). Pacific Grove, CA: Brooks/Cole.

Wu, C. H., Kao, J. C., & Chang, C. J. (1996). Analysis of outpatient referral failures. *Journal of Family Practice, 42,* 498–502.

Data and Documentation |

chapter 5

One of the challenges of my work was managing paperwork, particularly the case notes that were required after each client contact. Case notes began with the initial interview. I tried several ways to keep them current. The agency had no video equipment, and even if I had audio taped, there was no one to transcribe the tapes for the file. So I tried taking notes while the interviewee was present. Then I tried making notes after each interview. I noticed that one of my co-workers scheduled nothing after 4:30 so he could spend the last 30 minutes of the day updating his case notes. I tried this one afternoon but found that I got confused about who said what! It was apparent to me that it was in my best interest and that of the interviewee to write the case notes immediately after the interview while the information was fresh. TM

Here Tricia describes her experiences managing the data that she gathers as a helper and documenting the data in her agency records. She has learned that writing her case notes immediately after the interview helps her improve the accuracy of those notes. In this chapter you will read about the multiple data elements that you gather or review in the initial interview and some helpful guidelines for documenting your interactions with interviewees.

What Does Data Have to Do with the Initial Interview?

Data are integral components of the interviewing experience and represent the information related to the interviewee and the interviewing process. The interviewer may use previously gathered data to understand the client and her situation prior to the initial interview. The interviewer is also gathering data during the initial interview. In addition, the interviewer may recommend that additional data be obtained following the initial interview. Data elements might include information about interviewee's previous history in the social service or counseling arena and reports written by education, medical, psychological, or vocational specialists. There might be reports or histories written about home visits or interactions with family members and friends. All of these data are collected in a case file.

Does Every Interviewee Have a Case File?

Every interviewee has a file, a collection of information about that specific interviewee. The data can be present in the interviewee file prior to the interview, generated during the interview, or collected after the initial interview concludes. For instance, the interviewee may be a referral from another agency; that agency may also forward its files, providing a written record of its work with the interviewee. This file may include information related to services provided over time or may be forms related to the referral. Two examples of file contents from referral agencies follow.

EXAMPLE 1   Cervantes Jones, age 34, has been released from prison and reports to his parole officer for the first time. The data in his computer file represent a summary of his interface with the criminal justice system for the past sixteen years. It begins with a report from an arresting officer for armed robbery and ends with his court-ordered parole.

EXAMPLE 2   An eighth grader has been referred for a psychological evaluation. The psychologist conducts the initial interview after reviewing the student's records from the last two years.

Regardless of when the data are gathered, it becomes part of the interviewee's case file.

What Information Do I Need to Acquire to Help the Interviewee?

Basic information about the interviewee helps guide the helping process. Standard information includes name, address, contact information (telephone, mobile phone, e-mail), sex, marital status, race/ethnicity, and the presenting issue. Information gathered in the initial interview may be used to begin the case file. The interviewer gathers general information or data about the interviewee as well as specific information related to the nature of the service being requested. For instance, if the interviewee is seeking public housing, the data gathered might include (1) contact information about the interviewee, (2) information about the family, (3) housing history, (4) work history, and (5) financial history. For each presenting problem, the list of data or information needed is, in part, specific to the issues addressed. Review the following issues and related data needed to better understand and address the issues. This provides you with an understanding of the wide range of information that can be gathered about an interviewee.

POTENTIAL CHILD ABUSE (DALY, 1991)

- Child information (name and date of birth, sex)
- Guardian information (name, address, social security number, marital status, other children, other people in residence)
- Disclosure information (to whom child made allegations, statements, prior reports)
- Behavior/environment (names and contact information of school, church, and day care individuals with whom child has contact; and playmates and neighborhood contacts)
- Family issues or problems
- Alleged perpetrator (basic information, past contacts with child, last contact with child)
- Medical history
- Description of home environment
- Evidence
- Further child protection recommended
- Notes

ADULT REHABILITATION AFTER CATASTROPHIC ACCIDENT

- Name
- Address
- Contact information
- Family involved in care
- Physical assessment (multiple areas of care and summary)
- Psychological assessment
- Psychosocial assessment
- Social history
- Work history
- Prognosis for recovery
- Adjustment
- Work prognosis

MOBILE MEALS

- Name and relevant contact information
- Source of referral
- Directions to home
- Summary of need as described by the interviewee
- Criteria for care (Department of Health, n.d.)
  - Inability to carry out personal or domestic routines
  - Involvement in work, education, or learning cannot be sustained
  - Social support cannot be sustained
  - Family and other social roles cannot be undertaken

Where Do I Get All of the Information That I Need?

The first place to look for information is in the interviewee's case file. It includes information gathered by other agencies, other interviewers, case managers, and professional helpers. Reviewing this file helps the interviewer decide how to approach the initial interview.

During the initial interview, the interviewee will be a major source of information. The information that is shared goes into a new case file or an existing file. Other possible sources of information are parents, employers, teachers, school counselors, and previous service providers. In some cases it will be helpful to make referrals to gather addition information. Chapter 4 provides information about the why, when, and how of referral.

What Is a Medical Evaluation/Report?

A medical report provides the results of a physical examination by a general physician or a specialist in fields such as a cardiology, endocrinology, or oncology. This information may be necessary in order to explore the health status of the individual, the source of physical difficulties, or any physical limitations. The report may also clarify any discrepancies between or among sources. A standard medical report includes identifying information about the interviewee, tests and findings, assessment, and recommendations. Understanding medical evaluations and using the reports may be difficult without knowledge of medical terms. References such as the *Physician's Desk Reference* (PDR) will help you decode medical reports. Without medical expertise, the interviewer may have trouble evaluating the quality of the report.

What Is the Psychological Evaluation/Report?

The psychological evaluation is usually performed and the report written by a licensed psychologist to assess what is happening in an individual's life that interferes with positive interactions or feelings. The report format usually begins with identifying data about the interviewee, the reason for the referral, a review of the assessment procedures, background information about the individual and family, mental health status exam (observations about the individual), test results, and recommendations.

The focus of the evaluation and report depends upon the reason for the examination, and the recommendations target the issues presented in the purpose for the evaluation. In fact, the psychologist may ask you what questions you want answered. Let's look at the contents of a psychological evaluation/report in detail (Nail, 1997).

Psychological Evaluation

Name:

Case No.:

Date of Evaluation:

Date of Report:

Purpose of Evaluation: This section states the reason for the referral and may also include issues on which the professional chooses to focus. This is the place to "briefly introduce the patient and the problem" (Nail, 1997). It also organizes the report. For example, for each purpose of the report, there are relevant assessments, summaries, and recommendations.

Assessment: This section lists assessments and summarizes results.

Background Information: This information is restricted to what is relevant for the purpose of the evaluation. The background information is in chronological order and may include a brief history of psychological difficulties, help received, motivations, clarity of thought, and emotional state.

Mental Health Status Exam: This presents the professional's assessment of the individual's cognitive functioning including "knowledge-related ability, appearance, emotional mood, and speech and thought patterns" (Frey, 2006).

Results of the Evaluation: The examiner presents a hypothesis (i.e., individual has a mood disorder) and then provides evidence to support or refute this hypothesis.

Summary/Recommendations: An integration of all information gathered and recommendations based upon specific conclusions are here.

What is a Social History?

A social history is often taken by an interviewer, sometimes during the initial interview but more often in subsequent interviews. The social history provides information about the way "an individual experiences problems, past problem-solving behaviors, developmental stages, and interpersonal relationships" (Woodside & McClam, 2006). A carefully taken social history provides a complete picture of the individual's life history, rather than a record of episodes or highlights. It can include an assessment for services, as well as an evaluation of client strengths and problem-solving strategies. Care must be taken to establish the context of the social history within the context of the culture. One limitation of the social history includes the interviewee's focus on the past rather than the future.

An example of an intake social history that focuses on the presenting problem(s) illustrates one approach to taking a social history (University of Wyoming, 2007).

Intake Social History

Client:

Professional:

Identification number:

Date:

Identifying data: name, sex, birth date, race/ethnicity, living arrangements, family number, residence

Referral: Source of referral and reason for referral

Presenting problem: Discussion of the problem in the interviewee's own words, representing the interviewee perspective. Examples provided by the interviewee support perceptions. Interviewee provides this information:

a. What is the problem? An example? What would have to be done to solve the problem?
b. What is the history of the problem, including when the problem started and what life was like before the problem?
c. How often does the problem occur; how long does it last?
d. In what situation does the problem occur; what occurs to set off the problem; what thoughts or emotions occur before the problem?
e. How is the problem affecting the interviewee? Others?
f. What is the interviewee's goal in treatment?
g. What happens when the interviewee behaves appropriately or inappropriately (as defined by the interviewee)?
h. What are other problems related to presenting problem?
i. What are interviewee strengths? Needs?

Background information: This can include family configuration, abuse history, life development history, information about education, work, health, marriage, financial, legal. All information represents the interviewee's perspective.

How Do I Obtain a Release of Information?

There are several ways that you can gather additional information about the interviewee. Some only require you to inform the interviewee. Others require permission. One way to gather more information is to talk with others who currently have a relationship with the interviewee, such as family, friends, teachers, and other significant adults. In these instances, it is important for the interviewer to, at the very least, let the interviewee know that he intends to talk with these individuals. The interviewer can also ask for permission to talk with other individuals in the interviewee's world, understanding if the interviewee says no, then her wishes need to be respected.

You can request information from another source, such as a physician or the school, or you can make a referral for an additional professional assessment. Professional codes of ethics provide guidelines for these requests. For example, the American Counseling Association 2005 Code of Ethics states that permission must be obtained in writing from counselees (or guardians) to disclose or transfer information (American Counseling Association, 2005). There are also guidelines for special care when transmitting confidential information using electronic means such as e-mail, fax, and voice mail (American Counseling Association, 2005). A sample written permission for disclosure of information includes interviewee name, authorization, disclosure terms, information to be disclosed, and signature.

Obtaining/Release of Information

Name of Client

Date of Birth

I, _____, hereby authorize NAME OF SERVICES to obtain and release information pertaining to my evaluation and treatment to/from:

for the purpose of:

I understand that authorization is valid from the date of my signature below until:

I have been informed that I may revoke this authorization at any time orally or in writing to NAME OF SERVICES.

I certify that this form has been fully explained to me and I understand its contents.

Signature of Interviewee

Date of Signature

Signature of Interviewer

Date of Signature (Counseling and Psychological Services, University of Pennsylvania, n.d.)

 EXERCISE 5.1: **Release and Referral**

Video Clip A illustrates how one interviewer obtains a release and suggests a referral for testing. After you have watched this clip, answer these questions:

1. *How does the interviewer introduce the idea of gathering additional information?*
2. *What new information does the interviewer need? Why do you think the interviewer needs this information?*
3. *How does the interviewer explain the need for additional information to the interviewee?*
4. *How does the interviewer talk about a release of information?*

Why Do I Need to Provide Documentation of the Initial Interview?

Documentation is important for several reasons. First, your time with the interviewee becomes part of the case file; it affirms that the interviewee has begun interaction with you and your agency or organization. Whether your documentation is the first entry into the interviewee's case file or becomes part of a larger record, it may assist in providing continuity of care. For example, your documentation becomes part of the history of service delivered and provides information that is passed from caregiver to caregiver. In addition, documentation helps you think through what you have learned about the interviewee and what else you might need to know, and provides a structure within which you can summarize and make recommendations for interventions.

Documentation also provides a foundation for risk management (Patrick, 2007). If there is a question of quality of care, you have provided data that outline the nature of your interaction with the interviewee and a rationale for actions taken. A record of the interview can also become part of a larger database that measures presenting problems, amount of time spent with interviewees, and interventions. This information helps individual professionals and agencies evaluate outcomes and quality of the services provided.

How Do I Document What I Learned in the Initial Interview?

The written document that you provide about the initial interview becomes part of the interviewee's case file and is often called an *intake summary*. The way you document what you learned about the interviewee, in part, depends upon the purpose of the initial interview and whether the interview is structured or unstructured. For example, if you are using a structured intake form to guide the interview, you will add information, line by line, based upon the subject line in the interview form and the information the interviewee gives you. For example, a short version of a structured intake form used for youth intake or the initial interview at a runaway shelter follows.

Intake Interview

Personal Data

Name: Tom Smith

Date of birth: December 10, 20XX

Guardian: Mary Smith

Contact information: 999-9999 (mobile)

Educational Information

Name of school: Ross High School

List your educational goals. GED

What clubs or sports to you participate in? Football

If an interview is unstructured, then you write a summary of what the interviewee tells you. The next excerpt represents an unstructured approach to the school counselor's initial meeting with Tom.

Intake Interview Summary

Tom Smith is a 17-year-old single white male at Ross High School. He was referred to the school counselor by three teachers. He has missed more than 25 percent of his classes since fall break. His teachers and his mother are concerned about this change in behavior. Tom arrived promptly for our 10:30 appointment. He was neatly dressed and well groomed. He seemed a bit nervous at the beginning of the interview, shifting in his chair and crossing and uncrossing his arms.

After introducing myself and explaining my role as a school counselor, I asked Tom to tell me about himself and what he wanted to discuss with me.

As you conduct the unstructured interview, your follow-up questions will come from what the interviewee tells you. You'll probably want to make notes under various subheadings, and following the interview, write a summary of the information he provides.

Are There Any Guidelines to Follow When I Write My Notes About the Interview?

Professional documentation of the initial interview requires attention to detail and an accurate rending of the interaction and your observations. There are several guidelines that will help you provide a quality summary of your work.

- *Use simple, precise language.* Clear language provides a precise description of the initial interview. Avoid vague or general statements. For example:

VAGUE: The interviewee was unhappy.

PRECISE: The interviewee kept her head lowered during the interview; she did not make eye contact except when she entered the room and when she left the interview.

VAGUE: I do not think that interviewee wanted to be in the interview.

PRECISE: The interviewee spoke in monosyllables. He stated his mother made him come to the interview.

- *Write about important components of the interview.* Your observations about the interviewee are important. Consider including information about the interviewee's appearance, expression and gestures, change of emotion depending upon the subject matter mentioned, response to interventions or suggested interventions.
- *Quote with care.* It is important to convey what the interviewee has said, but if you quote, then only do so when you remember what the interviewee said. If you quote, you also need to describe the context of the statement. Most of your writing will, in all likelihood, be paraphrase.
- *Do not make judgments or disrespectful comments.* Words such as "dirty," "lazy," and "mean" are evaluative. These words are not helpful in a fair analysis of what the interviewee needs and may bias others who read the notes.
- *Write facts and not opinions.* Part of being clear about the interviewee and his or her behaviors, thoughts, and feelings is to distinguish between what you observed and what you think about what you observed. Remember to indicate when you are stating your impressions.

STATEMENT: The interviewee was late for the interview because it was not important to him.

IMPRESSION: The interviewee was late for the interview, and the interviewer thought it might be because the interview was not important to him.

STATEMENT: The interviewee said unkind things about her mother because her mother left her alone most of the day.

IMPRESSION: The interviewee said unkind things about her mother, and the interviewer hypothesized it was because her mother left her alone most of the day.

EXERCISE 5.2: **Record Keeping**

In Video Clip B, the interviewer discusses when and how she documents an interview. After you have watched the clip, do the following:

1. *List the tips suggested by the interviewer.*
2. *Watch Video Clip A again. Did the interviewer follow her own tips? Explain.*

COMMON BUT DIFFICULT: **Contradictory Information**

At the conclusion of an interview, the interviewer hopes to be able to summarize what he or she has learned, which is much like putting together the pieces of a puzzle. But sometimes the pieces do not seem to fit together; contradictory information is presented by the interviewee or by the interviewee and a significant other such as family, a teacher, or other helping professionals. One component of the interview summary may articulate relevant facts, sources of information, conclusions, and contradictions. Noting the inconsistencies assists in thinking about continued data gathering and recommending next steps with client.

> *Susie, a 16-year-old adolescent female, diagnosed with mild mental retardation and bipolar disorder, was referred to a residential treatment center because of violent outbursts against her foster mother and aggressive behavior toward her teacher in school. The helping professional who conducted her initial interview at the residential treatment center wrote, "Initially, she presented flat affect, made minimal eye contact with the [interviewer], and frequently looked down. Her demeanor was shy and withdrawn as she spoke in a quiet voice. At times, due to low volume or voice tone, I found it necessary to ask her to repeat verbalizations" (Crawford, 2008).*

The interviewer followed her description of how Susie presented herself with the statement noting the discrepancy between the reason for referral and initial presentation. She further indicated that Susie stated, "I need help with my anger at home" and recommended individual art therapy as a way to encourage Susie to express herself and her feelings.

CULTURAL CONTEXT: **The Issue of Respectful Language**

The way we refer to individuals reflects how we view them. Language may indicate respect or low opinion, value, or depersonalization. Historically individuals with disabilities were referred to by their handicapping condition, for instance, "the blind man" or "the crippled child" (Hanania, 1998). This meant that an individual was identified by his or her condition, encouraging stereotypes based upon one characteristic. Stereotyping makes it more difficult to view an individual as a complex person with multiple defining characteristics, including strengths and talents.

Our understanding of the power of language means we are careful as we describe interviewees in our written reports (Eichler & Burke, 2006). Preferred language for documentation of the interviewee with a disability is suggested by Rula Hanania (1998).

POOR	PREFERRED
the deaf	people who are deaf
the vision impaired	people with vision impairments
the disabled	people with disabilities
a victim of AIDS	a person with AIDS
victim	a person who has had or experienced a trauma
invalid	person who has a disability caused

ETHICAL CONSIDERATIONS: **Confidentiality and Documentation in a Managed Care World**

When you think about the initial interview, managed care may not immediately come to mind, but the information gained in this interview can often determine whether or not the individual receives services and the type of services he receives. Managed care exists to address rapidly rising health care costs, including mental health care, and is transforming treatments, goals, and outcomes (Patrick, 2007). Initial interviews become an entry point for funding decisions. One component of managed care decision making is a review of interviewee information to authorize treatment; many times this means providing written or verbal data to a managed care staff member.

What distinguishes this situation from providing information to another helping professional such as a physician, psychologist, social worker, or teacher is that many members of the managed care staff are not credentialed. While physicians, psychologists, social workers, and teachers are credentialed and bound by professional codes of ethics, particularly related to confidentiality (Woodside & McClam, 2006), managed care staff may not operate under such guidelines. How do interviewers resolve issues concerning confidentiality of interviewee information when communicating with managed care staff?

Guidelines will help you address this question.

1. If possible, interviewers need to understand, prior to the interview, if managed care organizations will receive information about the interviewee and how that particular organization handles the confidentiality of records.
2. The interviewer articulates to the interviewee the stated managed care policy and obtains permission to release the information on the interviewee's behalf.
3. Then the interviewer releases information based upon what he believes the managed care organization needs to know.

Personal Reflections

1. *Imagine you are the interviewee. Using the social history example in this chapter, complete a social history for yourself.*
2. *Observe a person or a situation for five minutes. Document your observations. Check your documentation by reviewing the guidelines in this chapter.*

REFERENCES

American Counseling Association. (2005). *ACA Code of Ethics*. Retrieved August 29, 2008, from http://www.counseling.org/Resources/CodeOfEthics/TP/Home/CT2.aspx

Counseling and Psychological Services, University of Pennsylvania. (n.d.). *Obtaining/releasing information form*. Retrieved August 29, 2008, from http://www.vpul.upenn.edu/caps/forms/Release%20of%20Information.doc

Crawford, C. S. (2008). *Art therapy with an adolescent: A case study*. Unpublished dissertation, Knoxville, TN.

Daly, L. (1991). The essentials of child abuse investigations and child interviews. *Institute for Psychological Therapies Journal, 3*. Retrieved August 29, 2008, from http://www.ipt-forensics.com/journal/volume3/j3_2_2.htm

Eichler, M., & Burke, M. A. (2006). The BIAS FREE Framework: A practical tool for eliminating social bias in health research. *Canadian Journal of Public Health*, 97(1), 63–68.

Frey, R. J. (2006). Mental health status examination. *Gale Encyclopedia of Medicine*. Retrieved on August 29, 2008, from http://www.healthatoz.com/healthatoz/Atoz/common/standard/transform.jsp?requestURI=/healthatoz/Atoz/ency/mental_status_examination.jsp

Hanania, R. (1998). *The language of disability: The University of Indiana*. Retrieved on August 30, 2008, from http://www.iidc.indiana.edu/cedir/language.html

Nail, G. (1997). *Some thoughts on the format for the psychological report*. Retrieved on August 29, 2008, from http://www.msresource.com/format.html

Patrick, P. K. S. (2007). *Contemporary issues in counseling*. Boston: Pearson.

United Kingdom, Department of Health (2008). *Publications, policy and guidance*. Retrieved on August 29, 2008, from http://www.dh.gov.uk/en/Publicationsandstatistics/Publications/PublicationsPolicyAndGuidance/DH_4009653

University of Wyoming. (2007). *Social Work methods I: Standardized client interview assignment*. Retrieved on August 29, 2008, from http://www.uwyo.edu/uwcc/Syllabi/Fall2007/scisowk3630.doc

Woodside, M., & McClam, T. (2006). *Generalist case management*. Pacific Grove, CA: Brooks/Cole.

ROLES AND RESPONSIBILITIES | CHAPTER 6

I worked in a program that served young children and their families. I had a family of my own and was going to school at the time, so I was committed but careful with my time. Most days I started work at 7:30 A.M., and ended the day around 5:30 P.M., after I had completed my paperwork. And, to be honest, I took many of my concerns home with me about the children and families with whom I worked. One colleague I worked with, who had the same caseload as I did, arrived at work about 8:30 A.M., walked around the building with a coffee cup in his hand and talked with folks until about 9:30 A.M. when he started his work with clients. He left the building promptly at 4:30 P.M. He saw his job as a job, but when he worked with children and families, he was very "on" and provided excellent services. A third colleague approached the job in yet a different way. She was extremely involved with the children and the families on her caseload. She spent evenings with the families, celebrated birthdays with the children she served, took the children to sports events during the evenings, and even visited families on the weekends. All three of us defined our work in very different ways, especially with regard to boundaries and how we constructed relationships with the children and their families. MW

As you have seen in the first five chapters, the roles and responsibilities of interviewers are varied. Different people perform them well in very different fashions, as Marianne's example shows. In Chapter 6 the focus is on being a novice interviewer, seeking supervision, avoiding or addressing burnout and promoting professional wellness, and recognizing ethical issues within the context of the interview.

What Are My Roles and Responsibilities?

Interviewing may be your primary responsibility, or it may be one responsibility among many. One place to begin identifying roles and responsibilities is in a job description. This is a written description that defines the work for which you are held accountable. Rarely, however, does a job description accurately reflect the actual work, so it may be more helpful to think about it as a place to begin defining your job. Agencies also have organizational charts, that is, documents or diagrams that represent the lines of authority and accountability. Your supervisor should be easily identifiable on this chart so you will know his or her place in the organization.

Talking with your supervisor will also clarify exactly what expectations exist for you as a member of the organization. For example, the job description for an agency case worker might read as follows:

Case Work Job Description

Provides care management and evaluations for involuntary hospitalization; interviews clients to obtain information including psychiatric history, mental status, social history, education and vocational background; conducts home visits to evaluate clients' need for services; documents client activity; participates in interdisciplinary team service delivery.

Questions that you might ask your supervisor to help you determine the realities of the job description are:

What are the goals of this department or unit?

How does interviewing fit into these goals?

Does this department have its own policies and procedures?

How do they relate to my position as interviewer?

How will I be evaluated?

Am I Really Going to be Able to Help Anyone with My Limited Skills?

We believe that it is important to understand the skills and experience that you have and to be aware of what you do not know. It is important to seek supervision if you are unsure of how to interview a client or sense that the interview is not going as well as it should. For example, an interviewee may be severely and persistently mentally ill, and your training may have focused on working with developmental difficulties and situational crises. If you cannot conduct the interview or if you are uncomfortable with the behavior the individual is exhibiting, you may want to seek support from another colleague during the interview. On the other hand, if you are able to conduct the interview but you are confronting a novel situation, such as talking with an elderly adult when you have worked only with children, then asking for supervision after the interview is appropriate. Within the helping professions, you have an ethical commitment to be aware of difficult situations and/or your professional limitations and to ask others for assistance (American Correctional Association, 1994; American Counseling Association, 2005).

What Do I Do if I Don't Know or Don't Understand Something?

The question of what you do and don't know builds on the previous question, "Am I really going to be able to help anyone with my limited skills?" It reflects a fear of uncertainty, a normal reaction for some beginning interviewers. Sometimes, though, this fear can be overwhelming for new interviewers, causing them to forget the very basic concepts and skills they have learned. Relying on empathy and acceptance and skills like listening and clarifying will help a new interviewer work through this fear. If the fear isn't recognized, then one consequence is that it may also cause us to view the interviewee as a label: the blind person, the manic depressive, or the welfare mother. Another reaction might be reluctance to talk about interview experiences and denial of fears and questions. Blaming the victim is another response as the interviewer focuses on interviewee reluctance and resistance. Having an available supervisor is critical for the new interviewer's professional growth, especially for addressing the fear and anxiety that may accompany meeting a specific responsibility for the first time.

How Can I Make the Most of Supervision?

Supervision has a number of benefits. One is support, which was just discussed. If you are new to interviewing, then you might have specific questions ("How long should I spend doing an initial interview?" "Where are the initial interview forms?") or need help with a particular interviewee ("My next interview is a referral from the shelter. I don't know anything about domestic violence."). You might also want help developing your skills ("How do you manage to get information without asking so many questions?" "I'd like to be as calm and confident as you are."). So a supervisor can be a teacher, a consultant, an advocate, as well as a supporter.

Now, you may be wondering how you can get the most from supervision. Suggestions follow to maximize your relationship with your supervisor.

- Identify your supervisor within your agency setting. If you do not have a supervisor, then identify an advanced professional who would be willing to mentor you.
- Work with your supervisor to determine the nature and scope of the supervision that he can provide.
- Establish a regularly scheduled time to meet with your supervisor.
- Use your time with your supervisor wisely by preparing for your meetings.
- Ask questions that relate to interviewing (e.g., goals and objectives of the initial interview, note taking and documentation, specific issues that arise in interview settings, building rapport, establishing expectations for the interview).
- Develop a plan for professional development.
- Understand ethical and legal issues that relate to the interviewing process.

What if I Make a Mistake?

Believe it or not, concern about making mistakes is common among beginning helpers, particularly interviewers who are often the first contact in an agency. In fact, it's one of the nagging questions that form a part of every helper's professional conduct and emotions, according to Kottler and Blair in their book, *The Imperfect Therapist* (1989). Even today, this is a recurring question from our students.

Often, we are our own worst enemies when we think about our own behavior. In some ways, we are harsher critics of ourselves than others are. So what if we made a serious mistake and it was publicly displayed? Our harsh critiques arise from our concern for appearances as well as a legitimate worry that a single error could destroy a career. At worst, results may be a reprimand from a supervisor or an inquiry from an ethics committee.

How do we minimize the potential for mistakes? The most important actions we can take keep risks under control. To do this, interviewers are careful to document, practice conservatively, and increase their self-awareness by monitoring their own emotional responses.

> Recently, a school counselor learned about a first grader with a history of sexual abuse. The child at age 6 is now a sexual predator. This counselor recognized that this was a complex situation that required expertise she did not have. She carefully documented her referral to a child abuse specialist but she didn't stop there. She was so upset about the details the grandparent had shared with her that she couldn't get it out of her mind. She finally decided she needed to share her reaction with the other school counselor.

This is an example of conservative practice. The school counselor recognized immediately that she wasn't competent to deal with this child and her concerns about other first graders were legitimate. She was also careful to document. On a personal level, she recognized that she was horrified about this victim's treatment from her parents and needed to deal with her own feelings. A rush to intervene on her part with this first grader would have been a tremendous mistake!

How Do I Know if I Have Helped Someone?

In some helping professions it's often difficult to know if you've made a difference. These programs differ from medicine, for example, where a physician listens to lungs or examines an artery. In most instances, there is a diagnosis, an intervention, and a measurement to see if the medical problems identified still exist, have improved, or have worsened. Although this is a simplistic description of the medical model of assessment, treatment, and re-assessment, there is no comparable test in the helping professions to find out what is wrong and how to fix it. Sometimes in an initial interview we are able to pinpoint the issues and suggest solutions. People are complex and, in most instances, so are their situations and problems. We interview them and refer or work with many of them; but our interaction is time limited. We may know we have made a positive difference in the short term, particularly if there is a behavior change, but rarely do we know what the long-term results are.

Some individuals we encounter in that initial interview do not complete the interview or do not return for services. Other individuals participate in the helping process, but end the process with unresolved issues. For a variety of reasons, we have not made a difference, and we find ourselves continually grappling with expectations such as "I will be able to help all of the individuals I interview," "If I have enough time, I will be able to help people change," and "I have a positive vision that I can share with those I interview." Positive expectations are important! Realistic expectations help you build the knowledge and skills you need to become an effective interviewer.

How Can I Keep the Issues That Interviewees Have from Affecting Me?

A recent development in the helping professions has been the focus on reactions of helpers to working with survivors of trauma (Sexton, 1999): "there is a cost to caring" (Figley, 1995, p. 1). Of course, the impact will depend on whom you interview. If you interview survivors of child abuse, natural disasters, violent crime, acts of genocide, or war trauma, you may begin to feel overwhelmed yourself. Attempts to identify these effects have coined terms such as *empathic stress, compassion fatigue, secondary traumatic stress*, and *vicarious traumatisation*. A new field, psychotraumatology, has emerged and is examining these constructs.

Individual reactions vary but may include distancing from another person or situation, numbing, grief, or voyeurism. Those with a history of personal trauma may be attracted to work with trauma survivors. Although experience with trauma may give some helping professionals greater sensitivity and insight, it may also make them more vulnerable to vicarious traumatization. For these helpers, seeking aid and support for themselves is critical. On an organizational level, those helping professionals who are affected may engage in more boundary violations, experience more job turnover, and have low morale. The organization may then incur the additional costs of recruiting and training new staff.

If you feel that interviewing survivors of trauma over and over begins to affect you, then there are several strategies to help. One is to recognize that these are normal responses to working with trauma victims. Actually, it might be of more concern if you are not affected in some way. If you are experiencing some effects, then it's important to recognize what is happening to you, know the signals of distress in yourself, and understand what your personal tolerance is for hearing traumatic experiences. Peer support, trauma training, supervision, and teamwork are all positive strategies to employ.

How Will I Know if I'm Burning Out?

At some time or another, feelings of frustration, boredom, or disappointment cause helpers to question what they do. Do you dread going to work? Are you bumping into furniture or tripping over the trashcan? Do you find that you are impatient with both clients and staff? Have you ever caught yourself referring to clients in a disparaging way? Do you wonder if you are accomplishing anything? These may be signs that you are burning out. It's important to recognize it and to take action.

In 1976, Christina Maslach defined burnout as "a syndrome of emotional exhaustion, depersonalization, and reduced personal accomplishment that can occur among individuals who do people work of some kind" (1976, p. 3). This is the general psychological stress of working with people that progresses gradually. Burnout occurs on two different levels. One is its effect on the individual. Indicators of burnout include gastrointestinal problems, substance abuse, exhaustion, sleep disturbance, inability to concentrate, and interpersonal difficulties. On an organizational level, observable symptoms are absenteeism, tardiness, and high job turnover. These are generally attributed to low salaries, an unsupportive administration, and difficulties in providing services.

What Can I Do to Prevent Burnout?

If you have identified burnout as an issue that you face in your work as an interviewer, there are several steps that you can take to begin to alleviate the syndrome. Changing your career orientation helps you find a realistic view of what work and career mean to you both personally and professionally. Specifically, this means making a realistic assessment of what you can do for those you interview and articulating the successful outcomes by which you can measure your progress or evaluate your work. For example, you may identify components of the job that are rewarding and giving yourself credit for successes instead of focusing on perceived failures.

Another approach to preventing burnout is to assess what coping strategies you use that are of limited value and to develop coping strategies and skills that promote positive mental and physical health. Negative coping strategies such as giving up, striking out at others, indulging yourself (addictions), blaming yourself, blaming others, and denial comprise characteristics of the burnout syndrome described earlier. Constructive coping can counter these negative responses and help you deal with perceived stress as an interviewer. Three approaches—appraisal-focused, problem-focused, and emotional-focused coping—provide options to prevent or address burnout (Weiten & Lloyd, 2006). Using Cognitive/Emotional/Behavioral Therapy to reduce catastrophic thinking, humor as a stress reducer, and positive reinterpretation are ways to help you think about the stress you are experiencing. Identifying specific problems and using a problem-solving approach, integrating time management into your work schedule, and improving self-control and self-discipline may help you address problems at work.

Improving the quality of life outside work also helps promote well-being. When work is your major or only focus, it must fill both psychological and social needs, such as the need to be close to others, to belong, to be creative, and to relax. Rarely can any one activity meet all of these needs. So take some time to relax, play, create, and be quiet during the week.

These Days Safety Seems to be a Real Concern. What Can I Do to Make Sure I'm Safe at Work?

Safety at work is a legitimate concern for everyone, but especially for those of us in helping professions. There are actions you can take at both the agency level and individual level that help you address these safety concerns.

Agency Level

Know the plan! Societal violence, aggression, and acting out behavior have increased in our society in recent decades, and many of the most publicized occurrences have been within education and agency settings (James & Gilliland, 2005; Thomas, 2006). Educators and human service professionals, clients, and bystanders are more at risk from violent behavior for a multitude of reasons, including bullying, family history, substance abuse, history of violence, mental illness, and social stressors, to name a few (James & Gilliland). Institutions are responding with what many are labeling the "safety plan" or the "crisis plan." These plans usually include precautions to insure the safety of professionals and those being served, as well as comprehensive procedures should aggression or violence occur. It is important that you know the plan at your work place and that you receive training related to your individual precautions as well as organizational precautions. A plan may include the following components:

- Legal aspects and responsibilities related to violence in the workplace
- Assessment of variables that may trigger violence
- Defusing techniques
- Agency procedures when early warning signs of violence exist
- Agency procedures when violence occurs

Individual Level

Much of what you will learn about violence in the workplace will be in organizational plans and training; however, we encourage you to take charge of your own education about your safety at work! We include some guidelines here, but also recommend James and Guilliland's *Crisis Intervention Strategies* and R. Murray Thomas's *Violence in America's Schools: Understanding, Prevention, Responses* as excellent resources.

Thinking about your own safety at work affects your preparation for the interviewee, your reactions to interviewee behavior, and reports of your interactions (James & Guilliland, 2005). As you prepare for an interviewee, keep in mind that there is always a potential of aggression or violence, especially if there are related factors such as substance abuse or a history of violence, or there are warnings from co-workers. Be sure that your office door is open, and you are not isolated with the interviewee. Do not behave in a way that is considered aggressive, for example, standing too close or making threatening movements. Keep your voice even and your tone measured. Especially with regard to a potentially violent client, don't enter a room before her; keep space between you and the client, and the exit; and make contingency plans for incidents that *might* occur. Keep a record of all interactions with interviewees. Be sure to include any indications of aggression or violence.

How Will I Know When I've Done All I Can Do to be Helpful?

In most cases, an initial interview is bounded by at least two factors that influence the ways that you can be helpful in the initial interview. First, the interview is usually time-limited. Whether you are conducting the interview in the office or at the interviewee's home, you have an idea about how much time you have to complete the interview. Second, the purpose of the interview frames the focus and the content of the interview. A mental health status assessment will be a very different initial interview from an application for financial support for health services. For example, if you are a school counselor or you are a mental health professional in a residential setting, the initial interview may be less than fifteen minutes, due to constraints of your schedule or the schedule of the interviewee. The interview may take place in the hallway or a recreational setting and may be focused on exploring an immediate need on the part of the interviewee. A home visit may take one or two hours, depending upon the number of individuals to be interviewed and the amount of information to be gathered.

Because of the limits of the initial interview, you may not be able to "do all that can be done" to help the interviewee. In fact, many of the interviewees you encounter will have multiple needs that cannot be addressed during your first encounter. Table 6.1 provides some realistic ways the initial interview might end and suggested

TABLE 6.1 | HAVE I DONE ALL I CAN DO? ISSUES AND RESPONSES

Issue	Response
Time Limitation	"I'll make a note here of the difficulties you mentioned so that we can address them next time."
Scope of Agency	"You mentioned two issues that our agency cannot help you with. I have some information about other community resources that can help you address these issues."
Need for Referral	"This initial interview is to gather basic information about you and your situation. At the conclusion of the interview, we will make plans to refer you to other professionals, if necessary."
Need for More Information	"I don't believe that I can make any recommendations for help until you and I gather further information. The next time we meet let's think about what we need and how to get it."
Input from Interviewee	"We've talked about a number of problems today. Which one is most important?"
Specific Next Steps	"We follow a certain process in interviewing and providing services. Let me outline those for you."

interviewer responses: referring, seeking additional information, and prescribing next steps.

How Will I Know When to End the Interview?

Every initial interview comes to an end. At this point, it is important to summarize what has occurred during the interview and briefly describe what happens next. The interviewer might simply state, "John, let's briefly review what we've talked about today and where we go from here." Or the interviewer may ask the interviewee to summarize: "Jan, today you've shared a lot of information with me about you and some challenges you're facing. I'm wondering what you think the most important points are."

Four recommendations will help you think about how to terminate an interview. First, let the interviewee know at the beginning of the interview about the time constraints and the purpose of the interview. Second, keep an eye on the clock and inform the interviewee when 10 minutes remain. Third, save time at the end of the interview to summarize with the interviewee and talk about next steps, including the request for additional information or referral. This may also be a time to ask for questions. Fourth, stand and usher the interviewee from the interview room, providing a signal that the interview is over. This closing is as important as the greeting discussed in Chapter 2.

Contributing to a positive feeling at the close are a handshake, reinforcement for coming, and a word of encouragement: "I'm glad we had a chance to meet and talk today. I've learned a lot about your situation, and I'm looking forward to working with you. Is the same time next week good for you?" If the interviewee will see someone else next time, you might say, "I'm glad you came in today. I think you will like Nancy Roth, the counselor you will see. As I told you, she is very knowledgeable about credit counseling, and I think you'll be pleased with the progress you will make with her."

Here's what Joe, a physical therapist who works with children with disabilities, says about concluding the initial interview:

I have learned from experience that the close is as important as the body of the interview. The first thing that I do is tell parents it is time to conclude the interview. The dialogue then changes immediately. Sometimes parents want to talk more and faster, as if they realize that they are about to lose their opportunity for help. I give them a piece of paper on a clip board to write down anything that they wish they had told me so we can discuss it during a follow-up visit. I also may have a few questions that I still would like to explore. I do this in a casual way.

I have some set activities in the close. I review information from the greeting about the agency, purpose of the interview, my credentials. I summarize what I have learned about them and their child and ask for feedback. Then I ask them to summarize for me what they have learned. In this part of the country, leave-taking is slow and neighborly. If the parents like to leave quickly, then I go at their pace. Sometimes one parent is ready to leave and the other would stay the day. I leave time sometimes for the "neighboring" that I equate with the small talk that occurs in the greeting. Providing the time for the close prevents feelings of abruptness for all of us.

What Are My Ethical and Legal Obligations as an Interviewer?

There is *no* substitute for knowing the laws of the state in which you practice. You must also know the rules and regulations of your employer. In addition, professional behaviors are expected to reflect ethical obligations established by professional codes of ethics. Values that ground an interviewer's ethical behavior are autonomy, non-maleficence, beneficence, justice, fidelity, and veracity. Here's what they mean in the initial interview:

• *Autonomy* Encouraging the interviewee to participate fully in the interview to determine who has access to information and to share whatever information he chooses.
• *Nonmaleficence* Keeping the interviewee physically, psychologically, and emotionally safe. If the interviewee becomes distraught or extremely stressed during the interview, the interviewer helps the interviewee de-stress before leaving the interview setting.
• *Beneficence* Acting in the interviewee's best interest by keeping her needs at the forefront.
• *Justice* Providing equal access for all interviewees, paying attention to issues of discrimination, and advocating for nondiscriminatory policies.
• *Fidelity* Attending to the trust the interviewee places in the interviewer. This trust is reflected in interviewee self-disclosure and the belief that the interviewer will advocate for the needs of the interviewee.
• *Veracity* Being honest with the interviewee, including truthfulness, professional disclosure, honesty about informed consent and confidentiality, and fair and honest feedback.

What Am I Legally Responsible to Tell Authorities about What an Interviewee Tells Me?

Confidentiality is fundamental to the helping relationship and restricts information the professional can share with others about the client. In our view this includes information provided to the interviewer in an initial interview. There are limits to confidentiality, however, such as client approval of release of information in the form of reports and assessments. Other limitations include court orders for information, state and federal statues, and the duty to warn and protect, including considerations when working with HIV-positive clients (Corey, Corey, & Callahan, 2007; Welfel, 2006).

Courts may order professionals to provide records and testimony when serving as experts or performing a mental status evaluation. Interviewers should inform those interviewed that information provided in the interview will be shared with the court (Cohen & Cohen, 1999). All states require reporting suspected child abuse to the appropriate public agency. Some states require helping professionals to report suspected elder abuse. In many states, criminal charges can be brought against the helper for failing to report. In spite of the state statues, many professionals do not disclose the abuse, fearing the break in confidentiality will hurt the therapeutic relationship (Kalichman, 1993). Four guidelines may help you when considering reporting suspected abuse.

1. *Awareness* Be aware of the statutes in your own state as well as the guidelines from your professional code of ethics and the policy of your agency related to reporting of child and elder abuse and neglect.
2. *Knowledge* Understand the signs that indicate possible abuse and neglect and how to conduct a risk assessment.
3. *Seek Supervision* In situations where you suspect abuse and neglect, seek the advice of supervisors and colleagues about how and when to take action.
4. *Document* Record the client behavior you have noted that indicates harm to others and the measures that you have taken to address this situation. This is important because it provides a history of the issue and describes how you, as the interviewer, met your obligation to the client and how and why you breached confidentiality.

What if I Think the Person Is Suicidal?

In each interview, an interviewer should be able to conduct a suicide risk assessment that helps guide the interviewer's subsequent action. An informal assessment allows the interviewer to determine the seriousness of risk.

The American Psychiatric Association suggests a ten-item screening device called *SAD PERSONS* to assess risk (Osterweil, 2007). Each item on SAD PERSONS is worth 1 point. In evaluating, 1 to 2 points indicates low risk, 3 to 5 points to moderate risk, and 7 to 10 designates high risk.

Sex (male)

Age less than 19 or greater than 45 years

Depression (patient admits to depression or decreased concentration, sleep, appetite, and/or libido)

Previous suicide attempt or psychiatric care

Excessive alcohol or drug use

Rational thinking loss: psychosis, organic brain syndrome

Separated, divorced, or widowed

Organized plan or serious attempt

No social support

Sickness, chronic disease

A helping professional has an obligation to protect the interviewee (ACA, 2005). Choices include asking the interviewee to share the information about self-harm with family and friends or breaking confidentiality to contact a crisis team or family or friends. The guidelines described related to "duty to warn" (awareness, knowledge, seek supervision, and document) are also relevant to protecting interviewees from self-harm.

What Are the Boundaries Between the Interviewee and Me?

Boundaries represent one dimension of the professional relationship between the interviewer and the interviewee. The interviewer is a professional who is committed to act in the best interest of the interviewee. Because of the power differential in the relationship, interviewees may not be able to define the boundaries of the helping relationship. It is the responsibility of the interviewer to be aware of boundary issues and respect the nature of the professional relationship. Will you share your home telephone number or cell number? Will you accept calls after hours? Will you meet clients after office hours or before? How available will you be? These are questions you will answer yourself or in consultation with your supervisor.

There are several areas where boundary issues remain less clear (College of Psychologists of Ontario, 1998): self-disclosure on the part of the interviewer; giving or receiving significant gifts; maintaining dual and overlapping relationships; becoming friends; conducting helping within a professional setting; careful use of touch.

Several examples help illustrate interviewers with boundary issues to consider.

- Tomorrow's interviewee is the son of her neighbor's best friend.
- An interviewee asks his interviewer for a job recommendation.
- The interviewer uses his hand and arm to guide the child he is interviewing to the small chair at the round table in his office.
- An interviewer meets interviewees with no transportation at a local restaurant in the interviewee's neighborhood.
- One interviewer shares personal information about herself to prompt self-disclosure from the interviewee.

These situations reflect real situations in which interviewers find themselves. Are the interviewer behaviors right or wrong? Sometimes, without more information it is difficult to tell. Interviewers might now agree on the "proper" way to conduct an interview or establish a relationship with the interviewee. The following guidelines will help you think about the boundary issues described above as well as any issues you might encounter as an interviewer (College of Psychologists of Ontario, 1998).

- Is this in the best interest of the interviewee?
- Am I serving my needs or the interviewee's needs?
- How does this behavior influence the outcomes of the interview?
- When should I consult with my supervisor about boundary issues?
- Is this interviewee getting different or "special" treatment from me?
- Am I taking advantage of the interviewee?
- Is my action in violation of my professional code or state or federal statue?

What Happens After the Initial Interview?

Actually, there are several scenarios that illustrate what follows the initial interview. One is that the interviewee goes on his or her way because the problem is solved, help is available elsewhere, or no help is desired or offered. Another scenario is referral for services either by another helper or at another place. And finally, you may continue to work with the interviewee. Each of these options builds on skills that have been introduced in this text. Let's examine them in more detail.

The first scenario describes situations where we only see an individual, couple, or family one time. This encounter may leave us wondering what has happened to them. One of our students worked in a school program for homeless children. Four siblings were present for several weeks; one day they were gone. The shelter reported that the mother and children left during the night for parts unknown. Our student was both puzzled and worried by this and, because she had no closure, spent a lot of time ruminating about what happened. Where did they go? Why did they leave? Were they okay? Should she have done something differently? These are questions with no answers. Seeking supervision and support from co-workers, engaging in activities that prevent burnout, and using other activities discussed in this chapter can help us deal with situations where there is no closure.

Another common scenario is referral. Needs and services often do not mesh between the interviewee and the agency, but assistance may be available elsewhere. So we use the referral skills in Chapter 4. A Grief Outreach Initiative serves children ages 6 to 16. Often, older family members like parents and grandparents also have unresolved issues related to grief and loss. Referral to local mental health centers, support groups, clergy, social workers, and counselors provides help to these sufferers. When and how to refer are discussed in Chapter 4. Referral skills may be useful during the initial interview or at any other time during the helping process.

The third scenario occurs when you and the interviewee continue to work together. Remember that communication is at the heart of the helping process. You've read about a number of communication skills that will continue to serve you well throughout the helping process. Listening, exploring, clarifying, questioning, and other helping skills are discussed in Chapter 4. Another facet of helping is considering the ethical and legal commitments that ground your work as a helper. You will continue to use many of the guidelines presented in this chapter as well as the professional code of ethics by which you are bound.

These examples illustrate ways that initial interviewing skills contribute to the entire helping process. So while the focus of this book and DVD is the initial interview, the skills you have read about and observed are skills that you will use throughout your work with people. Remember that the focus of your work is on providing help and assistance. You can seek supervision and consultation when challenges arise. We also encourage you to take care of yourself by respecting boundaries between the helper and the helpee, continuing to develop your knowledge and skills, and maintaining a balance between your professional and personal life.

EXERCISE 6.1: **Issues to Consider**

Watch 7 helping professionals share common but difficult situations they encounter in their work. For each situation think about what you would do.

1. *I worry about sending children home to a difficult situation.*
2. *Males are often hesitant to talk about problems.*
3. *What the interviewee needs isn't available in my community.*
4. *There are some interviewees with whom I have difficulty establishing rapport.*

Other challenging situations follow:

5. *My next interview is with Sudanese refugees. I know nothing about the Sudan.*
6. *My supervisor doesn't have the skills to help me.*
7. *A young woman shows up for the interview with a 3-year-old. Her babysitter couldn't make it at the last minute.*
8. *You work with victims of violent crime. Some days you dread going to work.*
9. *There is no policy for documentation in your agency.*
10. *Sometimes the goals of the agency conflict with what the interviewees need.*
11. *A child is the identified client, but the problem is in the home environment.*

CULTURAL CONTEXT: Interviewing Trauma Survivors

Among trauma survivors are refugees who have been forced from a location, usually their homes, to another location not of their choosing. Often categorized as "forced" or involuntary migrants, moving is due to political, religious, or ethnic persecution and/or war, and their departures are without preparation, plan, or choice (Pedersen & Carey, 2003). Eight commonly shared experiences of refugees are political repression, detention, torture, violence, disappearance of relatives, separation and loss (of both people and possessions), hardships, and exile. As a result of these kinds of experiences, many refugees are traumatized, uprooted, and find they must deal with a hierarchy of suffering. Shukri Sindi (*A Refugee from Iraq—Shukri Sindi,*), a young artist from Iraq, had the following experience:

> When he was 14, he and his nine brothers and sisters, as well as his parents, were forced to flee their home in northern Iraq. They left in the dead of the night for the safety of the Turkish border. Shukri's mother remembers waking her ten children: "Get your clothes on," she urged, "we're leaving." Shukri and his family are Kurds, an ethnic group whose rebellion has been crushed by Iraqi dictator Saddam Hussein after the Gulf War. There were rumors that Kurdish civilians were being gassed and that other chemical weapons had been used to annihilate women hanging laundry and children at play.

Traumatization occurs throughout the refugee experience. It results from loss, victimization, and adjustment. Many refugees find these experiences so difficult to cope with that psychological dysfunction results. Being forced to leave the familiar for the new and unfamiliar creates a sort of culture shock that requires adaptation. This uprooting is exacerbated by lack of support systems. Unfortunately, contributing to the trauma is the belief that some refugees have more right to be traumatized than others and, consequently, deserve services more than others.

> What happens when a family of twelve is uprooted? How do they cope without food or a place to sleep? During his three-year stay in refugee camps, Shukri remembers always being hungry and cold, and never being in school. Three years is a long time for a child to be away from school, but the Sindis had no choice in the matter. The family of twelve lived beneath two tents tied together. At first, the United Nations brought food and clothing, and sympathetic countries sent help as well. But as the crisis

dragged on, the international relief effort dwindled. There were reports of beatings and supplies being stolen by Turkish soldiers who controlled the camp. No one could leave the compound and protests over living conditions were put down immediately.

Interviewing refugees requires sensitivity, empathy, and understanding. A skillful interviewer will focus on the reasons for the migration, the availability of family and community support systems, similarity of new culture to the old, and the flexibility and adaptability of the family. Refugees will experience higher levels of stress than voluntary migrants, and so coping strategies and prior intercultural experiences are important. In addition, age and sex are factors in that younger refugees will adapt more quickly than older refugees and males more quickly than females. An assessment of this information assists in problem exploration and identification, and in determining next steps.

Interestingly, the refugee experience is also applicable to 9/11 survivors, particularly the experiences of violence, disappearance of relatives, separation and loss, hardships, and exile. Survivors of 9/11 recall watching the World Trade Centers collapse, seeing people jump, and knowing people who died. For these survivors traumatization is a reality, and many of the factors that are assessed for refugees would be important in this situation. For example, it would be important to know about support systems, flexibility and adaptation, and coping strategies.

CULTURAL CONTEXT: **How Do I Know When I Need Help?**

Being a healthy interviewer yourself is critical for the welfare of your interviewee. This means knowing not only when you need help but also getting help. Help may relate to professional knowledge and skills as well as professional and personal issues that interfere with acting in the best interest of the interviewee. Four scenarios illustrate situations when assistance is needed from others.

> Bud is a licensed mental health counselor who works with troubled teens with dual diagnosis. Bud abused drugs and alcohol as a teen, participated in a rehabilitation program, and until four months ago, had abused neither drugs nor alcohol for 10 years. His recent divorce has sent him into a tailspin. He has been able to manage his work responsibilities, but has begun drinking at night.
>
> Tonya is just tired. She is in an internship and is working 40 hours a week at a home for pregnant teens and 30 hours a week waiting tables so she can pay her living expenses. One of her responsibilities is to make home visits and interview teens who have been in the program and now are taking care of their babies at home. Her last four interviews have not gone well, and she has left each without completing the interview. Her supervisor talked to her today about being more patient and empathetic with these young mothers.
>
> Two of Susan's interviews today were with women who needed assistance while they were getting divorces. She is recently divorced herself and knows better than anyone how difficult it is to make ends meet. She doesn't need to hear it from these two women.
>
> Blair knows that she is in over her head. She works with homeless adults. She is frightened every time she drives into the inner city to the shelter. She can't understand what the interviewees say when she talks with them about the program. She does not understand how to establish rapport with them.

Each of these interviewees needs help with his or her work, for a variety of reasons. How do we identify when we need help? It is important to recognize and acknowledge situations that call for more knowledge and skills than we have, draw attention to our own unresolved issues, or cause us emotional pain or distress. Two behaviors that signal caution are communication problems that we are now experiencing, interpersonal difficulties that are new, or both. Help is available from co-workers, supervisors, and professional counselors.

REFERENCES

A Refugee from Iraq—Shukri Sindi. *Refugee USA: Meet New Americans.* Retrieved February 24, 2004, from http://www.refugeeusa.org/meet_amer/refugee_iraq.cfm

American Correctional Association. (1994). *ACA Code of Ethics.* Retrieved September 4, 2008, from http://www.aca.org/pastpresentfuture/ethics.asp

American Counseling Association (ACA). (2005). *ACA Code of Ethics.* Retrieved August 29, 2008, from www.counseling.org/Resources/CodeOfEthics/TP/Home/CT2.aspx

Cohen, E. D., & Cohen, G. S. (1999). *The virtuous therapist: Ethical practice of counseling and psychotherapy.* Pacific Grove, CA: Brooks/Cole.

College of Psychologists of Ontario (1998). Professional boundaries in health-care relationships. *The Bulletin, 25*(1). Retrieved September 23, 2008, from http://www.yoursocialworker.com/boundaries.htm

Corey, G., Corey, M. S., & Callahan, P. (2007). *Issues and ethics in the helping professions* (7th ed.). Pacific Grove, CA: Brooks/Cole.

Figley, C. R. (1995). Compassion fatigue as secondary traumatic stress disorder: An overview. In C. R. Figley (Ed.), *Compassion fatigue: Coping with secondary traumatic stress disorder in those who treat the traumatized* (pp. 1–19). New York: Brunner/Mazel.

James, R. K., & Gilliland, B. E. (2005). *Crisis intervention strategies.* Pacific Grove, CA: Brooks/Cole.

Kalichman, S. C. (1993). *Mandated reporting of suspected child abuse: Ethics, law and policy.* Washington, DC: American Psychological Association.

Kottler, J. A., & Blair, D. S. (1989). *The imperfect therapist.* San Francisco: Jossey-Bass.

Maslach, C. (1976). Burnout. *Human Behavior, 5*(9), 16–22.

NASW (1999). Code of ethics of National Association of Social Workers. Retrieved on September 23, 2008, from http://www.socialworkers.org/pubs/code/code.asp

Osterweil, N. (2007). APA: Simple screening improves suicide risk assessment. Retrieved September 23, 2008, from http://www.psychiatrictimes.com/display/article/10168/58341?pageNumber=3

Pedersen, P. B., & Carey, J. C. (2003). *Multicultural counseling in schools: A practical handbook.* Boston: Allyn & Bacon.

Sexton, L. (1999). Vicarious traumatisation of counselors and effects on their workplaces. *British Journal of Guidance and Counselling, 27*(3), 393–403.

Thomas, R. M. (2006). *Violence in America's schools: Understanding, prevention, responses.* Westport, CT: Praeger.

Weiten, W., & Lloyd, M. A. (2006). *Psychology applied to modern life: Adjustment in the 21st century.* Belmont, CA: Wadsworth.

Welfel, E. R. (2006). *Ethics in counseling and psychotherapy: Standards, research, and emerging issues* (3rd ed.). Pacific Grove, CA: Brooks/Cole.

Index

W

Wacinko, 41
Weifel, E. R., 66
Weiten, W., 12, 15, 62
Woodside, M., 25, 27, 41, 49, 55
"Working with Sign Language Interpreters in
    Human Service Settings" (Davis), 19
World Wide Web (WWW)
professional organizations online, 4
services sought via, 3
Wu, C. H., 41

Y

Yawkey, T. D., 28
Yellow Pages, services sought via, 2